Table of Contents

FRACTIONS

Pre-test 2

Review 3

Naming fractional parts of a whole 4

Naming fractional parts of a set 10

Counting, ordering and comparing fractions 14

Adding and subtracting fractions with common denominators 19

Adding and subtracting mixed numbers with common denominators 26

Post-test 30

GEOMETRY

Pre-test 31

Point, line, line segment 32

Ray, angle 34

Line position: horizontal, vertical, diagonal 35

Parallel lines, intersecting lines, perpendicular lines 36

Line of symmetry, congruent figures 37

Solids: cube, cone, sphere, cylinder 40

Post-test 42

MEASUREMENT

Pre-test 43

Telling time to 15 and 5 minutes 44

Reading scales: temperature, weight, calendar 46

Measuring length to the nearest 1/2 inch 48

Converting and estimating length 51

Measuring capacity and weight in the customary system 53

Measuring length in the metric system .55

Estimating and converting lengths in the metric system 56

Measuring capacity and weight in the metric system 58

Finding perimeter and area 61

Adding, subtracting and multiplying with money 68

Pictographs and bar graphs 72

Probability 76

Post-test 80

Name _____ Score _____

Fractions Pre-Test: Level B

30. What fraction is shaded? _____

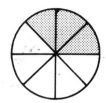

31. What part of the set is circled? _____

32. Write the correct sign: $>$, $<$, $=$,

a.

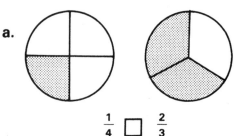

$$\frac{1}{4} \ \square \ \frac{2}{3}$$

b.

$$\frac{2}{4} \ \square \ \frac{1}{2}$$

33.a.
$$\frac{2}{8}$$
$$+ \ \frac{3}{8}$$
_____

b. $\frac{7}{9} - \frac{5}{9} =$ _____

34.a.
$$3\frac{3}{5}$$
$$+ \ 2\frac{1}{5}$$
_____

b. $8\frac{4}{7} - 6\frac{2}{7} =$ _____

1. How many equal parts?

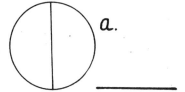 a. _____ 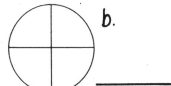 b. _____ 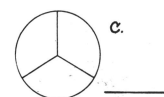 c. _____

2. Draw a broken line to divide each figure into...

a. $\frac{1}{4}$'s    b. $\frac{1}{3}$'s [   ]   c. $\frac{1}{2}$'s [   ]

3. What fractional part is shaded?

 a. _____   b. _____ 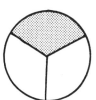 c. _____

 d. _____   e. _____   f. _____

Match each fraction to its word name.

_____ 4. one half   a. $\frac{3}{4}$

_____ 5. one fourth   b. $\frac{1}{3}$

_____ 6. one third   c. $\frac{1}{4}$

_____ 7. two thirds   d. $\frac{1}{2}$

_____ 8. three fourths e. $\frac{2}{3}$

Fair parts in a fraction

A fraction is a whole divided into fair parts.

Which bar shows fair cuts for the children?

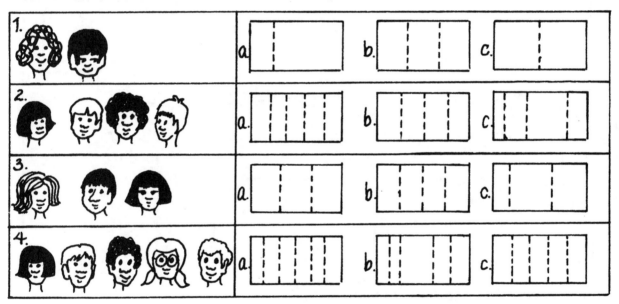

Draw a dotted line to show fair parts for the children.

Whole - Part - Fraction - Word Chart

1 whole	divided into a number of fair parts	1 part as a fraction	is called in words
	2 parts	$\frac{1}{2}$	one-half
	3 parts	$\frac{1}{3}$	one-third
	4 parts	$\frac{1}{4}$	one-fourth
	5 parts	$\frac{1}{5}$	one-fifth
	6 parts	$\frac{1}{6}$	one-sixth
	8 parts	$\frac{1}{8}$	one-eighth
	10 parts	$\frac{1}{10}$	one-tenth

EXERCISES

Write the shaded fraction in numbers and in words.

1. _____ _____

2. _____ _____

3. _____ _____

4. _____ _____

5. _____ _____

6. _____ _____

7. _____ _____

8. _____ _____

Write the fraction and word names for each shaded part.

1.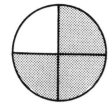

$$\frac{3}{4}$$

three-fourths

2.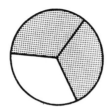

_____

_____

3.

_____

_____

4.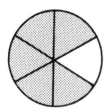

_____

_____

5.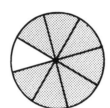

_____

_____

6.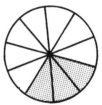

_____

_____

7.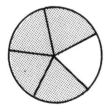

_____

_____

8.

_____

_____

9.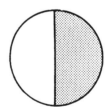

_____

_____

UniFix cubes of two different colors can show fractions.

What fraction is shaded?

1. _____

2. _____

3. _____

4. _____

5. _____

6. _____

7. _____

8. _____

TEACHER NOTE: Use Unifix cubes of two colors.
(1) "Make a 'candy bar' of two colors with the cubes."
(2) "How many cubes or parts does your candy bar have?"
(3) "How many of your cubes are the color _____?"
(4) "What fractional part of the bar is the color _____?"
    "Write the fraction on your paper and show it to me."
(5) "What does the fraction you have written mean?"
    (e. g. 2/7 would mean 2 of the 7 equal parts are yellow)

What part is shaded?

1.

$\left(\dfrac{1}{2}\right)$ $\dfrac{1}{3}$ $\dfrac{1}{4}$

2.

$\dfrac{1}{2}$ $\dfrac{3}{2}$ $\dfrac{3}{5}$

3.

$\dfrac{1}{3}$ $\dfrac{1}{2}$ $\dfrac{1}{4}$

4.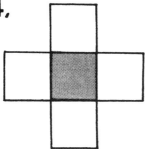

$\dfrac{1}{4}$ $\dfrac{1}{5}$ $\dfrac{4}{5}$

5.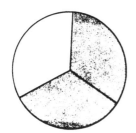

$\dfrac{1}{2}$ $\dfrac{3}{8}$ $\dfrac{3}{5}$

6.

$\dfrac{3}{4}$ $\dfrac{5}{7}$ $\dfrac{5}{6}$

7.

_____

8.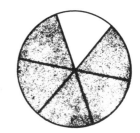

_____

9.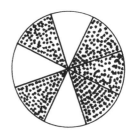

_____

10.

_____

11.

_____

12.

_____

Fractions as part of a set

EXERCISES

1. What part of the set is shaded?

_____

2. What part of the set is shaded?

_____

3. What part is circled?

_____

4. What part is circled?

_____

5. What part is circled?

_____

6. What part is circled?

_____

7 What part is circled?

_____

8. What part is circled?

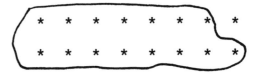

_____

Here is a set of children.

1. How many children in the total set? _____
2. How many children are wearing glasses? _____
3. What fraction of the children are wearing glasses? _____
4. What fraction of the children are not wearing glasses? _____

What fraction of each set is shaded?

5.

6.

7.

_____

8.

9.

10.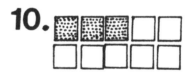

_____

TEACHER NOTE: Students write answer on individual chalkboards or on paper and show to the teacher after each question.

(1) "How many students are in the whole room? What fraction of the class are you? Write down the fraction.

(2) Whole numbers usually stand for one thing at a time. Fractions stand for two things at once. What two things does the fraction you have just written stand for?

(3) What fraction of the class are girls. Write down the fraction.

(4) What fraction of the class are boys (or have tennis shoes on or are wearing red, etc)? Write down the fraction and show it to me.

(5) Look at how many people are in your row (or table). What fraction of the row are you? Write it down and show it to me.

(6) What fraction of the class are you? What fraction of your row are you? These two fractions are different, but they both represent you. Can anyone explain how two different fractions can represent the same thing?"

Complete the table for sets 1 to 10 above.

Set	Shaded Objects	Objects in the set	Fraction of set shaded
1			
2			
3			
4			
5			
6			
7			
8			
9			
10			

Shade a part of the set for each fraction.

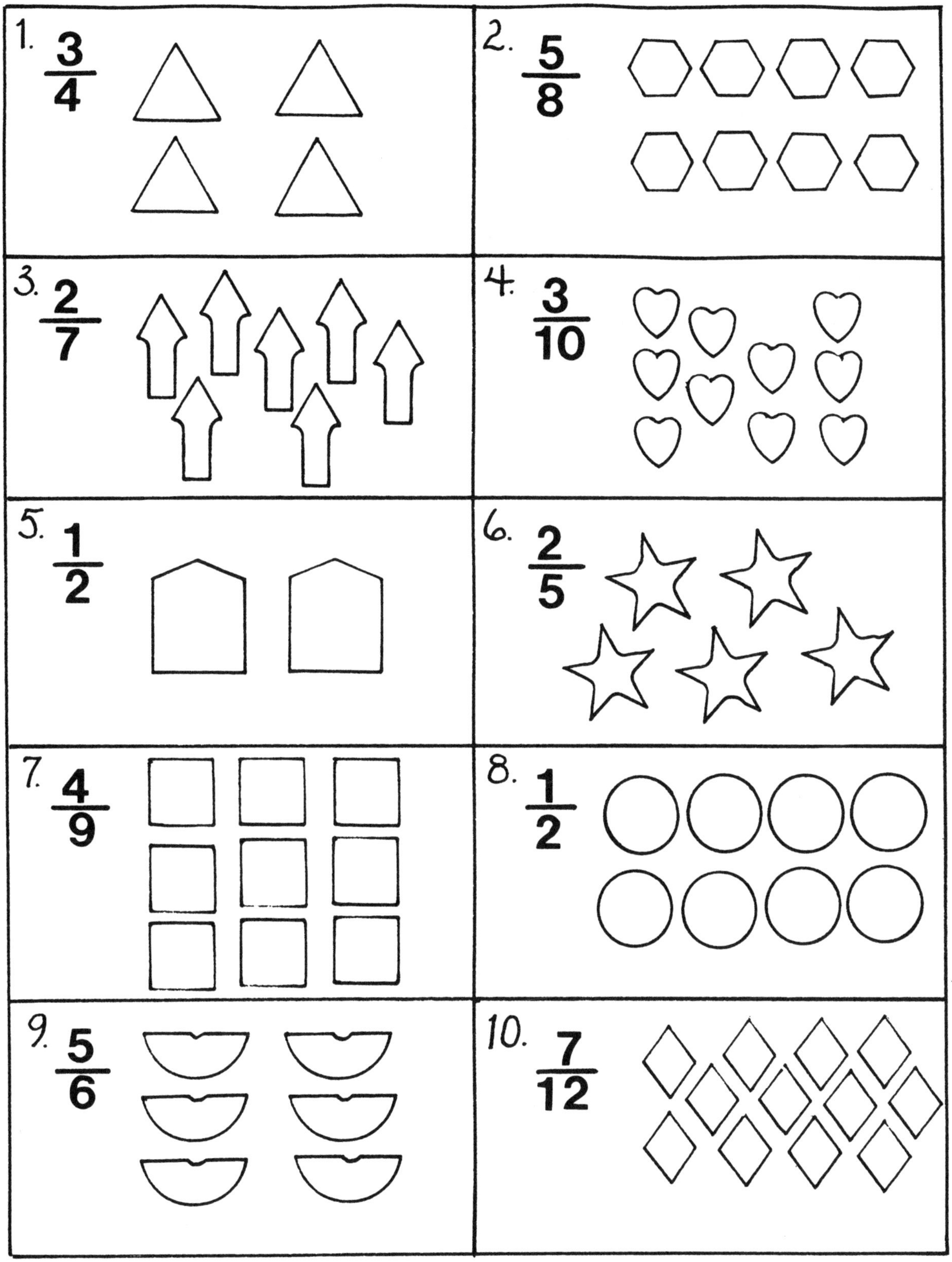

1. $\dfrac{3}{4}$

2. $\dfrac{5}{8}$

3. $\dfrac{2}{7}$

4. $\dfrac{3}{10}$

5. $\dfrac{1}{2}$

6. $\dfrac{2}{5}$

7. $\dfrac{4}{9}$

8. $\dfrac{1}{2}$

9. $\dfrac{5}{6}$

10. $\dfrac{7}{12}$

Counting fractions

Money can help with counting fractions to one.

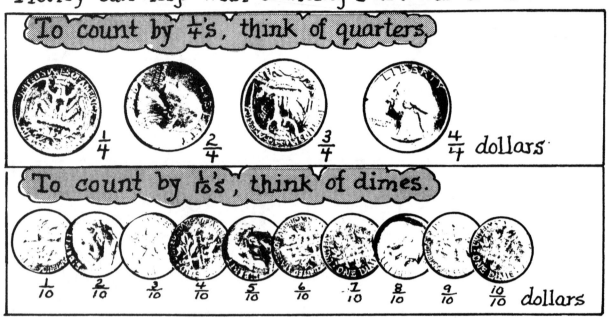

1. Count by $\frac{1}{8}$'s from $\frac{1}{8}$ to $\frac{8}{8}$:

___ , ___ , ___ , ___ , ___ , ___ , ___ , ___

2. Count by $\frac{1}{5}$'s from $\frac{1}{5}$ to $\frac{5}{5}$:

___ , ___ , ___ , ___ , ___

3. Count by $\frac{1}{6}$'s from $\frac{1}{6}$ to $\frac{6}{6}$:

___ , ___ , ___ , ___ , ___ , ___

4. Count by $\frac{1}{3}$'s from $\frac{1}{3}$ to $\frac{3}{3}$:

___ , ___ , ___

5. Count by $\frac{1}{9}$'s from $\frac{1}{9}$ to $\frac{9}{9}$:

___ , ___ , ___ , ___ , ___ , ___ , ___ , ___ , ___

Ordering fractions

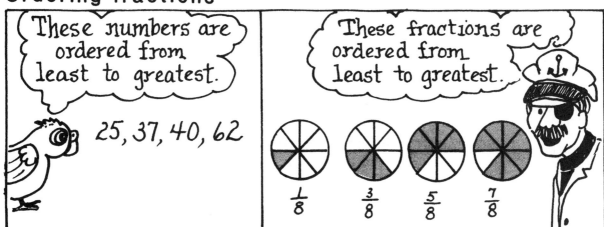

EXERCISES

Order from least to greatest.

1. 56   62   60   58

____   ____   ____   ____

2. 458   450   404   485   480

____   ____   ____   ____

3. $\frac{3}{4}$ $\frac{4}{4}$ $\frac{1}{4}$ $\frac{2}{4}$ $\frac{0}{4}$

4. $\frac{3}{5}$ $\frac{1}{5}$ $\frac{2}{5}$ $\frac{5}{5}$ $\frac{0}{5}$

____   ____   ____   ____   ____

5. $\frac{6}{6}$ $\frac{4}{6}$ $\frac{2}{6}$ $\frac{1}{6}$ $\frac{5}{6}$

6. $\frac{3}{10}$ $\frac{8}{10}$ $\frac{0}{10}$ $\frac{5}{10}$ $\frac{1}{10}$

____   ____   ____   ____   ____

7. $\frac{3}{12}$ $\frac{5}{12}$ 1 $\frac{2}{12}$ $\frac{1}{12}$

8. $\frac{4}{9}$ $\frac{2}{9}$ $\frac{8}{9}$ $\frac{1}{9}$ $\frac{3}{9}$

____   ____   ____   ____   ____

9. $\frac{1}{2}$ $\frac{0}{4}$ $\frac{1}{4}$ $\frac{2}{2}$ $\frac{3}{4}$

10. $\frac{1}{2}$ $\frac{1}{8}$ $\frac{3}{8}$ $\frac{0}{8}$ $\frac{1}{4}$

____   ____   ____   ____   ____

Comparing fractions with the same number of matching parts.

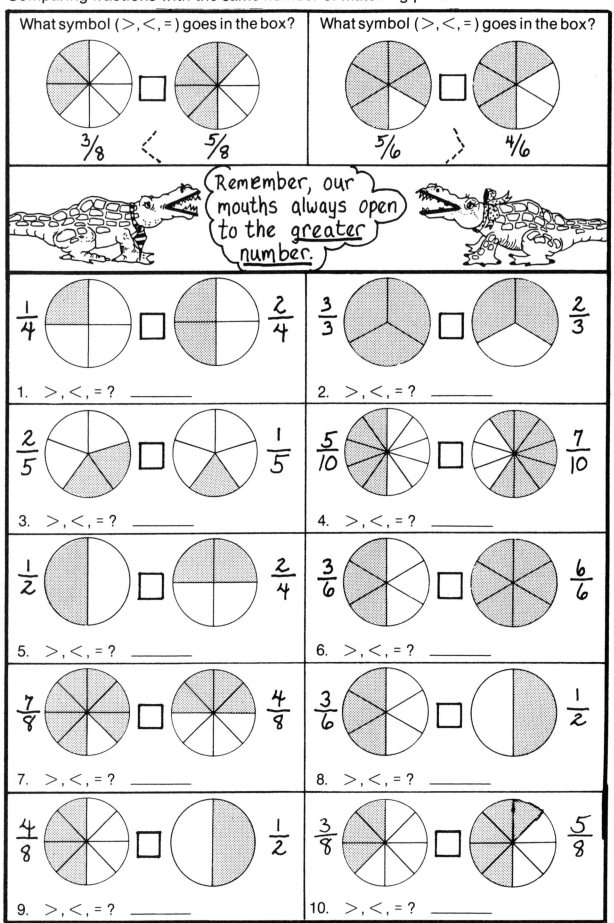

What symbol (>, <, =) goes in the box?

$\frac{3}{8}$ ☐ $\frac{5}{8}$

What symbol (>, <, =) goes in the box?

$\frac{5}{6}$ ☐ $\frac{4}{6}$

Remember, our mouths always open to the <u>greater</u> <u>number.</u>

$\frac{1}{4}$ ☐ $\frac{2}{4}$

1. >, <, = ? _____

$\frac{3}{3}$ ☐ $\frac{2}{3}$

2. >, <, = ? _____

$\frac{2}{5}$ ☐ $\frac{1}{5}$

3. >, <, = ? _____

$\frac{5}{10}$ ☐ $\frac{7}{10}$

4. >, <, = ? _____

$\frac{1}{2}$ ☐ $\frac{2}{4}$

5. >, <, = ? _____

$\frac{3}{6}$ ☐ $\frac{6}{6}$

6. >, <, = ? _____

$\frac{7}{8}$ ☐ $\frac{4}{8}$

7. >, <, = ? _____

$\frac{3}{6}$ ☐ $\frac{1}{2}$

8. >, <, = ? _____

$\frac{4}{8}$ ☐ $\frac{1}{2}$

9. >, <, = ? _____

$\frac{3}{8}$ ☐ $\frac{5}{8}$

10. >, <, = ? _____

Use the strips to compare the fractions.

	1				
	$\frac{1}{2}$		$\frac{1}{2}$		
	$\frac{1}{3}$	$\frac{1}{3}$		$\frac{1}{3}$	
$\frac{1}{4}$	$\frac{1}{4}$		$\frac{1}{4}$		$\frac{1}{4}$
$\frac{1}{5}$	$\frac{1}{5}$	$\frac{1}{5}$	$\frac{1}{5}$		$\frac{1}{5}$

1. $\frac{1}{4} \bigcirc \frac{1}{2}$

2. $\frac{1}{2} \bigcirc \frac{1}{3}$

3. $\frac{1}{3} \bigcirc \frac{1}{5}$

4. $\frac{1}{2} \bigcirc \frac{1}{5}$

5. $\frac{1}{3} \bigcirc \frac{1}{4}$

6. $\frac{1}{5} \bigcirc \frac{1}{4}$

7. $\frac{1}{2} \bigcirc \frac{2}{3}$

8. $\frac{3}{4} \bigcirc \frac{1}{2}$

9. $\frac{2}{5} \bigcirc \frac{1}{2}$

10. $\frac{2}{5} \bigcirc \frac{2}{3}$

11. $\frac{2}{4} \bigcirc \frac{2}{5}$

12. $\frac{2}{4} \bigcirc \frac{1}{2}$

13. $1 \bigcirc \frac{1}{2}$

14. $\frac{2}{3} \bigcirc 1$

15. $1 \bigcirc \frac{3}{4}$

16. $\frac{5}{5} \bigcirc 1$

REMEMBER...
"Our mouths always open to the GREATER NUMBER!"

Fill the circle with $<$, $>$, or $=$.

1. $\frac{1}{8} \bigcirc \frac{1}{2}$

2. $\frac{1}{2} \bigcirc \frac{7}{10}$

3. $\frac{5}{6} \bigcirc \frac{1}{2}$

4. $\frac{1}{2} \bigcirc \frac{4}{8}$

5. $\frac{2}{3} \bigcirc \frac{1}{2}$

6. $\frac{5}{10} \bigcirc \frac{1}{2}$

7. $\frac{1}{4} \bigcirc \frac{2}{3}$

8. $\frac{1}{2} \bigcirc \frac{3}{6}$

9. $\frac{3}{4} \bigcirc \frac{1}{3}$

10. $\frac{1}{2} \bigcirc \frac{1}{3}$

11. $\frac{3}{8} \bigcirc \frac{3}{4}$

12. $\frac{4}{5} \bigcirc \frac{3}{10}$

13. $\frac{1}{4} \bigcirc \frac{2}{3}$

14. $\frac{3}{10} \bigcirc \frac{3}{4}$

15. $\frac{5}{6} \bigcirc \frac{2}{5}$

16. $\frac{4}{10} \bigcirc \frac{6}{8}$

17. $\frac{2}{3} \bigcirc \frac{2}{6}$

18. $\frac{1}{10} \bigcirc \frac{3}{5}$

19. $\frac{2}{3} \bigcirc \frac{2}{6}$

20. $\frac{5}{8} \bigcirc \frac{1}{3}$

$$\frac{1}{4} + \frac{2}{4} = ?$$

Fraction circles can help you add fractions.

Step 1. Build each of the fractions to be added.

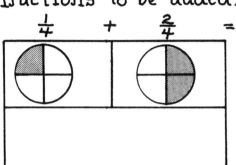

$$\frac{1}{4} + \frac{2}{4} =$$

Step 2. Put the fractions together to show addition.

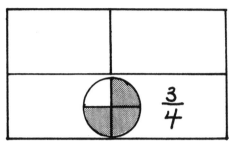

$$\frac{3}{4}$$

Step 3. Say the problem: "One-fourth plus two-fourths equal three-fourths."

Write each problem in words.

1. $\frac{1}{4} + \frac{1}{4}$ = _____ _____

2. $\frac{1}{3} + \frac{1}{3}$ = _____ _____

3. $\frac{2}{5} + \frac{1}{5}$ = _____ _____

4. $\frac{3}{8} + \frac{2}{8}$ = _____ _____

TEACHER NOTE:

(1) "How many students are in the whole room? How many are in the row nearest the door? What fraction of the whole class is the row nearest the door? Write it down and show it to me."

(2) "How many students are in the second row from the door? What fraction of the whole class is in the second row from the door? Write it down and show it to me."

(3) "If we added the fraction for the people in the first row and the people in the second row, what fraction of the class would we have? Write it down."

(4) "Can you write these three fractions in an equation? An equation must have an equals sign. Show me your equation. Say the fraction word names."

(5) Repeat steps 1-4 with other examples. Ask students to look for a pattern in the answers they have written down.

(6) Write several problems, e.g. 2/5 + 1/5 on the chalkboard. Ask students to write the answers. Some may write 3/5 and some may write 3/10; compare these answers to the pattern developed in steps 1-4 above.

Fraction problems can be written two ways.

① $\frac{5}{9} + \frac{2}{9} = \frac{7}{9}$ or

Add the top numbers

Keep the **bottom numbers**

②
$$\frac{5}{9}$$
$$+\frac{2}{9}$$
$$\frac{7}{9}$$

EXERCISES

1. $\frac{1}{3}$
$+\frac{1}{3}$

2. $\frac{2}{4}$
$+\frac{1}{4}$

3. $\frac{2}{8}$
$+\frac{1}{8}$

4. $\frac{2}{6}$
$+\frac{3}{6}$

5. $\frac{1}{5}$
$\frac{1}{5}$
$+\frac{2}{5}$

6. $\frac{1}{8}$
$\frac{2}{8}$
$+\frac{4}{8}$

7. $\frac{3}{10}$
$\frac{2}{10}$
$+\frac{4}{10}$

8. $\frac{2}{7}$
$\frac{3}{7}$
$+\frac{1}{7}$

9. $\frac{4}{8} + \frac{1}{8} =$ _____

10. $\frac{2}{5} + \frac{2}{5} =$ _____

11. $\frac{5}{10} + \frac{4}{10} =$ _____

12. $\frac{4}{9} + \frac{1}{9} + \frac{2}{9} =$ ___

13. $\frac{7}{12} + \frac{1}{12} + \frac{3}{12} =$ ___

14. $\frac{4}{11} + \frac{2}{11} + \frac{3}{11} =$ ___

15. Kevin mowed $\frac{1}{4}$ of the lawn before dinner and $\frac{2}{4}$ of the lawn after dinner. What fraction of the lawn did he mow? _____

Subtracting fractions

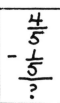

$$\frac{4}{5}$$
$$-\frac{1}{5}$$
$$\overline{?}$$

Fraction circles can help you subtract fractions.

Step 1. Build the greater number.

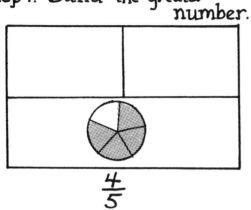

$$\frac{4}{5}$$

Step 2. Remove the lesser number.

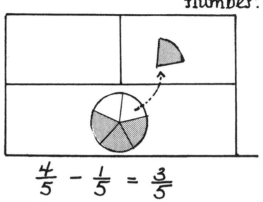

$$\frac{4}{5} - \frac{1}{5} = \frac{3}{5}$$

Step 3. Say the problem: "Four-fifths minus one-fifth equals three-fifths."

Use your mat to help solve these problems.

1. $\frac{3}{4}$
    $-\frac{2}{4}$
    _____

2. $\frac{5}{6}$
    $-\frac{4}{6}$
    _____

3. $\frac{7}{8}$
    $-\frac{2}{8}$
    _____

4. $\frac{1}{2}$
    $-\frac{1}{2}$
    _____

5. $\frac{2}{3}$
    $-\frac{1}{3}$
    _____

6. $\frac{11}{15}$
    $-\frac{3}{15}$
    _____

7. $\frac{4}{13}$
    $-\frac{2}{13}$
    _____

8. $\frac{6}{7}$
    $-\frac{5}{7}$
    _____

TEACHER NOTE: Students are to write the answers to each question on a chalkboard or paper and show it to the teacher.

(1) "What fraction of the whole class is boys? Write it down and show it to me."

(2) "I am going to ask two boys, _____ and _____, to stand outside the doorway. What fraction of the class is standing outside the doorway?

(3) What fraction of the boys are still left in the classroom?

(4) Write the three fraction in an equation. Show . . . Say the fraction word names."

(5) Repeat steps 1-4 with other examples. Ask students to find the pattern.

(6) Write several problems, e.g. 5/7 - 2/7, on the chalkboard for students to work. Compare answers to the pattern developed in steps 1-4 above.

Each of these circles has one-half of it shaded.
These shaded parts are other names for one-half.

$\frac{1}{2}$	$\frac{2}{4}$	$\frac{3}{6}$	$\frac{4}{8}$	$\frac{5}{10}$
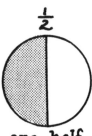	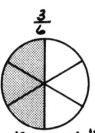	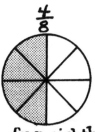	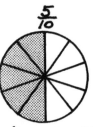	
one-half	two-fourths	three-sixths	four-eighths	five-tenths

The simplest name for all these fractions is $\frac{1}{2}$.

1. Which of these fraction circles have one-half shaded?

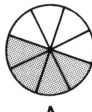 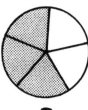 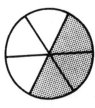  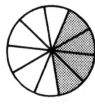

   A   B   C   D   E

2. Which of these are other names for $\frac{1}{2}$?

a) $\frac{2}{4}$ b) $\frac{2}{3}$ c) $\frac{3}{5}$ d) $\frac{4}{8}$ e) $\frac{5}{10}$

Add or subtract. Write the answers in simplest form.

3. $\frac{1}{4} + \frac{1}{4} =$ 4. $\frac{1}{6} + \frac{2}{6} =$ 5. $\frac{3}{10} + \frac{2}{10} =$

6. $\frac{5}{8} - \frac{1}{8} =$ 7. $\frac{5}{6} - \frac{2}{6} =$ 8. $\frac{7}{8} - \frac{3}{8} =$

A. How many different names can you write for $\frac{1}{2}$?

B. Can you find a rule to help decide if a fraction is equal to $\frac{1}{2}$?

Other names for one

Each of these circles has all of its parts shaded.
Each of these circles are other names for 1.

$1 = \frac{2}{2} = 1$  $1 = \frac{4}{4} = 1$ $1 = \frac{8}{8} = 1$ $1 = \frac{6}{6} = 1$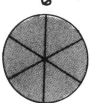

The simplest name for all these fractions is 1.

1. Count by $\frac{1}{5}$'s from $\frac{1}{5}$ to $\frac{5}{5}$ or 1.

____ , ____ , ____ , ____ , ____ or ____ .

2. Count by $\frac{1}{8}$'s from $\frac{1}{8}$ to $\frac{8}{8}$ or 1.

____ , ____ , ____ , ____ , ____ , ____ , ____ , ____ or ____ .

3. Count by $\frac{1}{6}$'s from $\frac{1}{6}$ to $\frac{6}{6}$ or 1.

____ , ____ , ____ , ____ , ____ , ____ or ____ .

4. Count by $\frac{1}{10}$'s from $\frac{1}{10}$ to $\frac{10}{10}$ or 1.

____ , ____ , ____ , ____ , ____ , ____ , ____ , ____ , ____ , ____ or ____ .

5. Which of these are other names for 1 ?

a.) $\frac{7}{7}$ b.) $\frac{1}{7}$ c.) $\frac{8}{9}$ d.) $\frac{5}{5}$ e.) $\frac{12}{12}$ f.) $\frac{1}{1}$

g.) $\frac{1}{4}$ h.) $\frac{10}{10}$ i.) $\frac{2}{2}$ j.) $\frac{5}{1}$ k.) $\frac{3}{3}$ l.) $\frac{12}{1}$

A. How many different fraction names can you write for the number 1?

B. Can you find a rule to help decide whether a fraction is equal to 1 ?

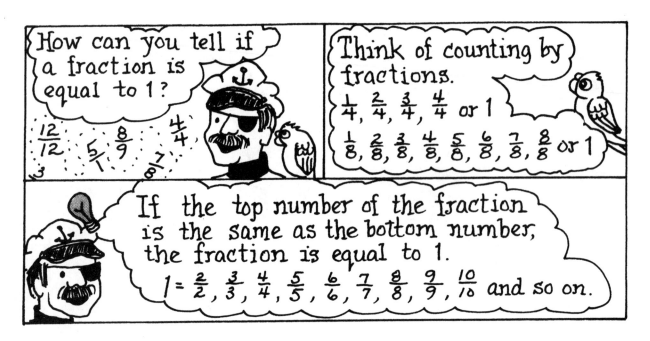

1. Draw four shaded fractions equal to 1.

2. Write five fraction names equal to 1.

___ ___ ___ ___ ___

What number goes in the box?

3. $1 = \dfrac{\boxed{}}{4}$ 4. $1 = \dfrac{\boxed{}}{5}$ 5. $1 = \dfrac{\boxed{}}{10}$ 6. $1 = \dfrac{\boxed{}}{8}$

7. $\dfrac{3}{3} = \boxed{}$ 8. $\dfrac{12}{12} = \boxed{}$ 9. $\dfrac{2}{2} = \boxed{}$ 10. $\dfrac{6}{6} = \boxed{}$

11. $1 = \dfrac{3}{\boxed{}}$ 12. $1 = \dfrac{7}{\boxed{}}$ 13. $1 = \dfrac{8}{\boxed{}}$ 14. $1 = \dfrac{9}{\boxed{}}$

Add or subtract these fractions. Change the answers to the simplest form.

1. $\frac{1}{2}$
   $+\frac{1}{2}$
   _____

2. $\frac{3}{4}$
   $+\frac{1}{4}$
   _____

3. $\frac{2}{5}$
   $+\frac{3}{5}$
   _____

4. $\frac{1}{6}$
   $+\frac{5}{6}$
   _____

5. $\frac{4}{7}$
   $+\frac{3}{7}$
   _____

6. $\frac{1}{10}$
   $+\frac{9}{10}$
   _____

7. $\frac{2}{3}$
   $+\frac{1}{3}$
   _____

8. $\frac{5}{8}$
   $+\frac{3}{8}$
   _____

9. $\frac{1}{12}$
   $+\frac{11}{12}$
   _____

10. $\frac{4}{9}$
    $+\frac{5}{9}$
    _____

11. $\frac{5}{11}$
    $+\frac{6}{11}$
    _____

12. $\frac{11}{15}$
    $+\frac{4}{15}$
    _____

13. $\frac{9}{8}$
    $-\frac{1}{8}$
    _____

14. $\frac{3}{2}$
    $-\frac{1}{2}$
    _____

15. $\frac{7}{4}$
    $-\frac{3}{4}$
    _____

16. $\frac{7}{5}$
    $-\frac{2}{5}$
    _____

17. $\frac{13}{10}$
    $-\frac{3}{10}$
    _____

18. $\frac{9}{6}$
    $-\frac{3}{6}$
    _____

19. $\frac{15}{12}$
    $-\frac{3}{12}$
    _____

20. $\frac{11}{6}$
    $-\frac{5}{6}$
    _____

21. The fourth grade held a cake sale. Each cake was divided into 2 equal parts. Sara sold $\frac{1}{2}$ of a cake to Mrs. Bryd and $\frac{1}{2}$ of a cake to Mr. Dean. What part of a cake did Sara sell? _____

Adding whole numbers and fractions

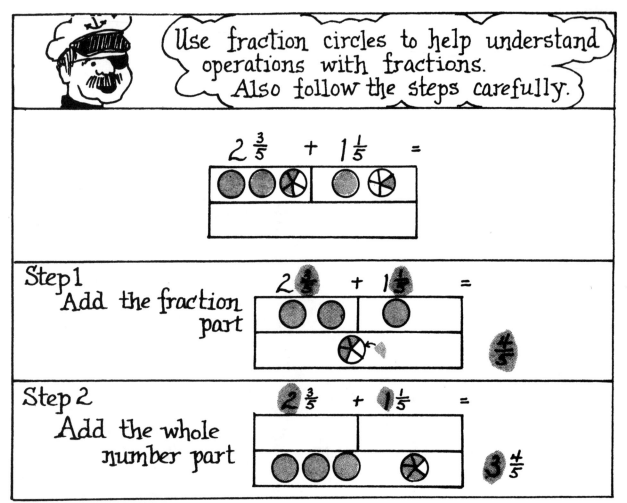

Use fraction circles to help understand operations with fractions.
Also follow the steps carefully.

$2\frac{3}{5} + 1\frac{1}{5} =$

Step 1
Add the fraction part

$2\frac{3}{5} + 1\frac{1}{5} =$ $\frac{4}{5}$

Step 2
Add the whole number part

$2\frac{3}{5} + 1\frac{1}{5} =$ $3\frac{4}{5}$

EXERCISES

Add.

1. $2\frac{2}{4}$
   $+2\frac{1}{4}$

2. $3\frac{2}{5}$
   $+2\frac{2}{5}$

3. $1\frac{1}{3}$
   $+3\frac{1}{3}$

4. $1\frac{1}{8}$
   $+1\frac{2}{8}$

5. $1\frac{2}{10}$
   $+2\frac{5}{10}$

6. $2\frac{4}{6}$
   $+1\frac{1}{6}$

7. $1\frac{5}{8}$
   $+3\frac{2}{8}$

8. $2\frac{4}{7}$
   $+2\frac{2}{7}$

9. Connie walked $1\frac{7}{10}$ mile on Saturday and $1\frac{2}{10}$ mile on Sunday. How many miles did Connie walk?

Subtracting whole numbers and fractions

Fraction circles can help you to understand subtracting whole numbers and fractions. $3\frac{3}{4} - 1\frac{2}{4} = ?$ 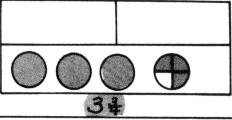	Step 1. Build the greater number on your mat. $3\frac{3}{4}$
Step 2. Remove the fraction part of the lesser number. 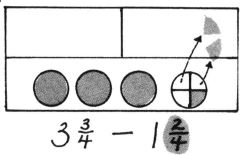 $3\frac{3}{4} - 1\frac{2}{4}$	Step 3. Remove the whole number part of the lesser number. 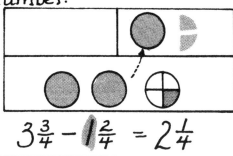 $3\frac{3}{4} - 1\frac{2}{4} = 2\frac{1}{4}$

Subtract.

1. $2\frac{3}{4}$
    $-1\frac{2}{4}$

2. $3\frac{2}{3}$
    $-2\frac{1}{3}$

3. $2\frac{3}{5}$
    $-2\frac{1}{5}$

4. $3\frac{6}{8}$
    $-1\frac{3}{8}$

5. $2\frac{7}{10}$
    $-2\frac{4}{10}$

6. $3\frac{1}{2}$
    $-1\frac{1}{2}$

7. $3\frac{5}{6}$
    $-1\frac{4}{6}$

8. $3\frac{6}{7}$
    $-2\frac{3}{7}$

9. Nick had a piece of canvas $4\frac{2}{3}$ yards long. He cut off $2\frac{1}{3}$ yards to make an apron. How much canvas does Nick have left?

 TEACHER NOTE: To avoid cutting the fraction circles into fractional parts which may get lost, you may wish to have students cover the parts to be removed with a kernel of corn. The answer is the number not covered.

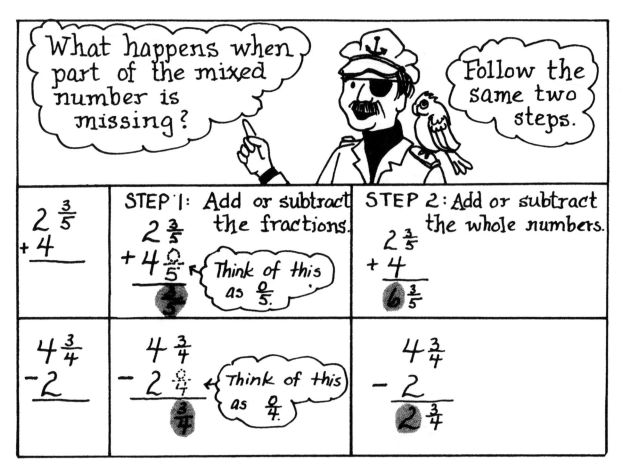

1. $2\frac{2}{3}$ 2. $1\frac{3}{10}$ 3. $1\frac{3}{4}$ 4. $2\frac{3}{8}$
   $+\ 1$ $+\ 1$ $+\ 2$ $+\ 2$
   _____   _____   _____   _____

5. $4\frac{4}{12}$ 6. $8\frac{3}{7}$ 7. $7\frac{5}{16}$ 8. $9\frac{7}{11}$
   $-\ 1$ $-\ 2$ $-\ 3$ $-\ 5$
   _____   _____   _____   _____

9. Han worked 3 hours on Monday and $2\frac{3}{4}$ hours on Tuesday. How many hours did Han work?

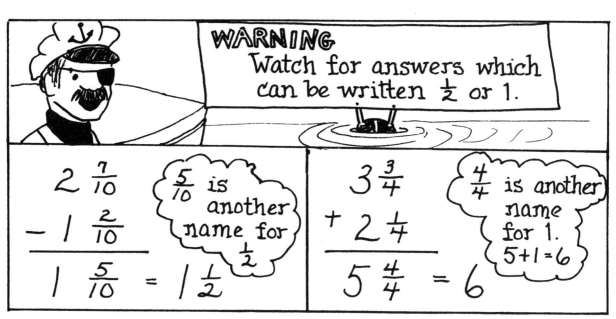

WARNING
Watch for answers which
can be written $\frac{1}{2}$ or 1.

$2\frac{7}{10}$
$-1\frac{2}{10}$
_____
$1\frac{5}{10} = 1\frac{1}{2}$

$\frac{5}{10}$ is another name for $\frac{1}{2}$

$3\frac{3}{4}$
$+2\frac{1}{4}$
_____
$5\frac{4}{4} = 6$

$\frac{4}{4}$ is another name for 1.
$5+1=6$

Add or subtract. Write the answer in simplest form.

1. $1\frac{1}{2}$
   $+2\frac{1}{2}$
   _____

2. $2\frac{3}{4}$
   $+3\frac{1}{4}$
   _____

3. $4\frac{3}{8}$
   $+2\frac{5}{8}$
   _____

4. $6\frac{7}{10}$
   $+1\frac{3}{10}$
   _____

5. $2\frac{1}{6}$
   $+5\frac{2}{6}$
   _____

6. $4\frac{5}{12}$
   $+3\frac{1}{12}$
   _____

7. $2\frac{1}{4}$
   $+3\frac{1}{4}$
   _____

8. $3\frac{1}{8}$
   $+5\frac{3}{8}$
   _____

9. $6\frac{7}{10}$
   $-3\frac{2}{10}$
   _____

10. $5\frac{6}{8}$
   $-3\frac{2}{8}$
   _____

11. $9\frac{7}{12}$
   $-2\frac{1}{12}$
   _____

12. $7\frac{3}{4}$
   $-5\frac{1}{4}$
   _____

Name _____ Score _____

Fractions Post-Test: Level B

30. What fraction is shaded? _____

31. What part of the set is shaded? _____

32. Write the correct sign: $>$, $<$, $=$,

a.
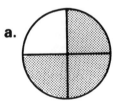 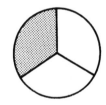

$$\frac{3}{4} \; \square \; \frac{1}{3}$$

b.
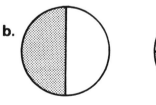 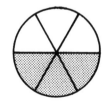

$$\frac{1}{2} \; \square \; \frac{3}{6}$$

33. a. $\quad \frac{3}{10} + \frac{4}{10} =$ _____

b.
$$\begin{array}{r} \frac{4}{5} \\ - \frac{2}{5} \\ \hline \end{array}$$

34. a.
$$\begin{array}{r} 2\frac{1}{3} \\ + 4\frac{1}{3} \\ \hline \end{array}$$

b. $6\frac{7}{8} - 2\frac{4}{8} =$ _____

Geometry Pre-Test: Level B

35.

What is the name of this figure?

_____

36. Line _____ is vertical.

(A)   (B)

(C)   (D)

37. Lines _____ are parallel.

(A)    (B)

(C) 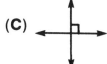   (D)

38. Which shows a line of symmetry? ____

(A)   (B)

(C)   (D)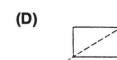

39. Which figures are congruent? _____

(A)   (B)

(C)   (D)

40.

What is the name of this solid?

_____

Geometry: point, line, line segment

Use your pencil to trace over these figures from geometry. Learn how to label and name them.

A point is shown with a dot and a capital letter.

P.

This is point P.

A line is a straight path that goes on and on in both directions.

This is line AB or BA or $\overleftrightarrow{AB}$. Notice the arrows at both ends of the line.

←————•————————•————→
    A   B

A line segment is a part of a line.

This is line segment AB or BA or $\overline{AB}$. Notice the two end points at A and B.

•————————————•
A   B

Complete.

1.   •————————•
    C   D

    Name this figure in two ways. How many end points does it have?

    _____

2.   ←————————→
     C   D

    Name this figure in two ways. How many end points does it have?

    _____

3. Draw a figure showing $\overleftrightarrow{XY}$.

4. Draw a figure showing $\overline{XY}$.

5. How is $\overleftrightarrow{XY}$ different from $\overline{XY}$?

    _____

6. Which is longer, a line or line segment? _____

A. Draw a point X on your paper. Draw 3 lines passing through point X. How many lines can you draw through point X?

B. Draw a point X and Y on your paper. Draw a line passing through X and Y. How many lines can you draw through X and Y?

C. How many points are needed to determine a line or line segment?

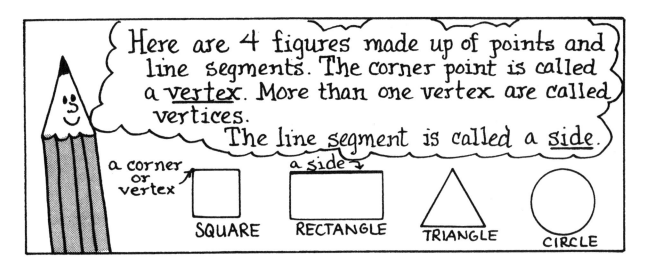

Here are 4 figures made up of points and line segments. The corner point is called a **vertex**. More than one vertex are called vertices.

The line segment is called a **side**.

a corner or vertex

a side

SQUARE RECTANGLE TRIANGLE CIRCLE

1. Which of the figures above is made up of a curved line? ____ ___ __

2. How many line segments or sides are there in a square? _____

3. A rectangle has how many sides? _____ _____

   How many vertices? _____

4. A triangle has how many sides? ___

   How many vertices? _____

How many vertices ? How many sides?

5.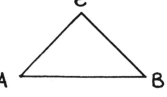

6.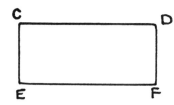

7.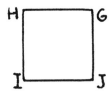

8.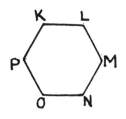

Count and name the line segments.

A   B   C   D

Ray, angle

| A ray is a part of a line.  | An angle is two rays having the same end point.

 The common end point is the vertex. 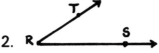 |

Complete.

1.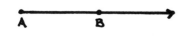

   What is this figure called? How many end points does it have?

   _____

2.

   What is this figure called? What is the vertex?

   _____

3. Draw a figure showing $\vec{CD}$.

4. Draw a figure showing $\vec{CD}$ and $\vec{CE}$ having a common end point C.

5. Name this figure: _____

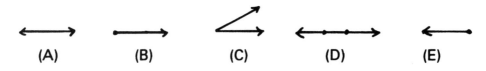

6. Name this figure in three ways:

   _____ _____ _____

7. Which of the following are rays? _____

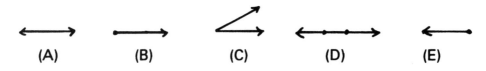

   (A)   (B)   (C)   (D)   (E)

8. Which of the following are angles? _____

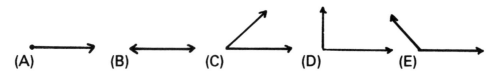

   (A)   (B)   (C)   (D)   (E)

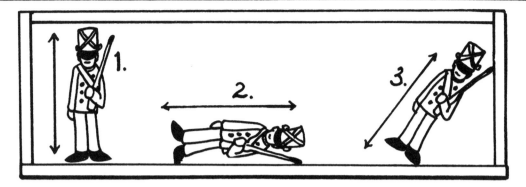

The toy soldiers are in different positions on the shelf.
Soldier #1, standing straight up and down, is in vertical position.
Soldier #2, lying across the shelf, is in horizontal position.
Soldier #3, leaning against the shelf, is in a slanting or diagonal position.

What is the position of each line?

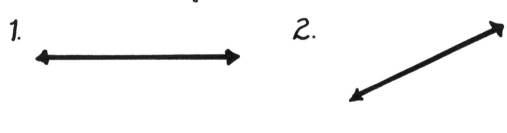

1.

2.

3.

4.

TEACHER NOTE: Have students use geoboards and rubber bands to demonstrate vertical, horizontal and diagonal lines.

Parallel lines, Intersecting lines

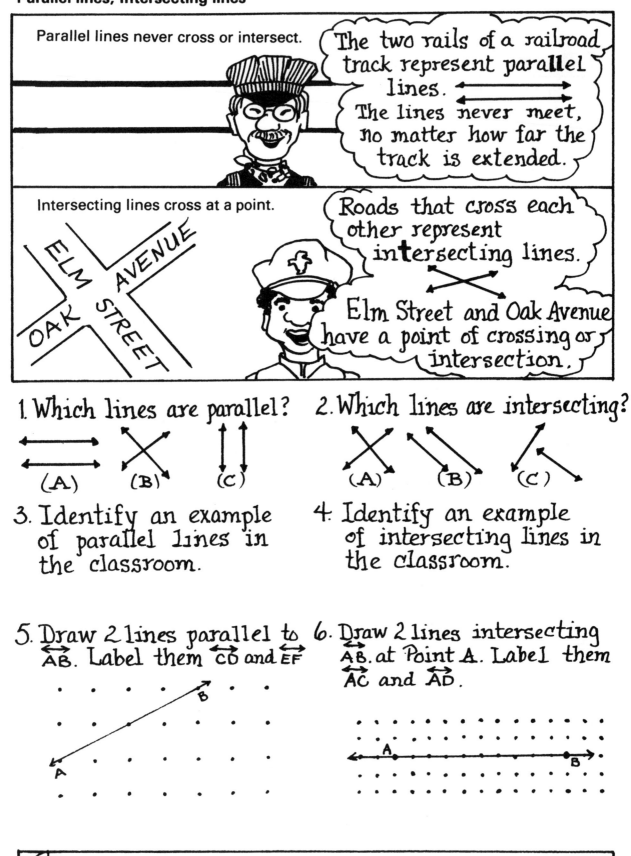

Parallel lines never cross or intersect.

The two rails of a railroad track represent parallel lines.
The lines never meet, no matter how far the track is extended.

Intersecting lines cross at a point.

Roads that cross each other represent intersecting lines.

Elm Street and Oak Avenue have a point of crossing or intersection.

1. Which lines are parallel?

(A)   (B)   (C)

2. Which lines are intersecting?

(A)   (B)   (C)

3. Identify an example of parallel lines in the classroom.

4. Identify an example of intersecting lines in the classroom.

5. Draw 2 lines parallel to $\overleftrightarrow{AB}$. Label them $\overleftrightarrow{CD}$ and $\overleftrightarrow{EF}$

6. Draw 2 lines intersecting $\overleftrightarrow{AB}$ at Point A. Label them $\overleftrightarrow{AC}$ and $\overleftrightarrow{AD}$.

TEACHER NOTE: Have teachers use geoboards and rubber bands to make one line parallel to another and one line intersecting another.

Perpendicular lines

Intersecting lines which meet at square angles are called perpendicular lines.

The corners of many rooms and most picture frames are made of perpendicular lines. 	These lines are perpendicular to each other. 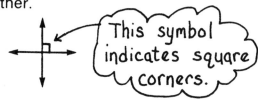

1. Which lines are perpendicular? _____

(A) (B) (C) (D)
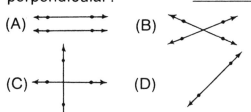

2. Which lines meet at square corners? _____

(A) (B) (C) (D)
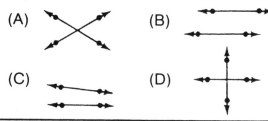

3. Draw a line perpendicular to line AB.

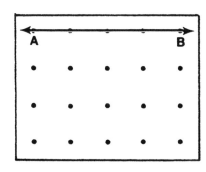

4. Draw a line perpendicular to line AB.

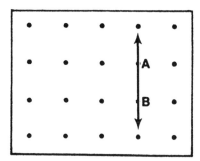

5. Draw a line perpendicular to XY.

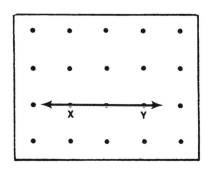

6. Draw a line perpendicular to line XY.

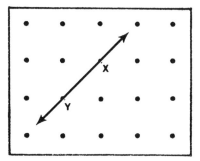

Find examples of perpendicular lines about you.

Line of symmetry

When you fold on a line of symmetry, the two parts should match.

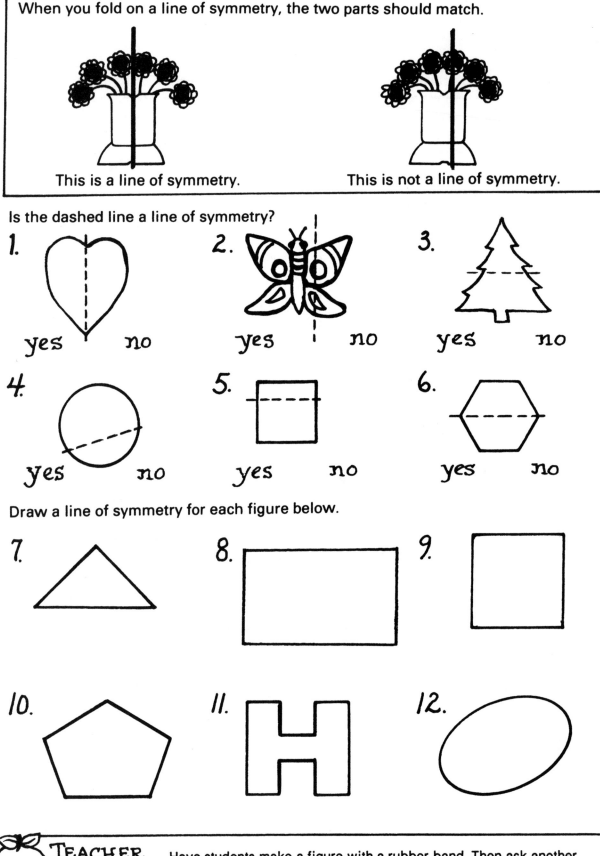

This is a line of symmetry.   This is not a line of symmetry.

Is the dashed line a line of symmetry?

1.   yes   no

2.   yes   no

3.   yes   no

4.   yes   no

5.   yes   no

6.   yes   no

Draw a line of symmetry for each figure below.

7.

8.

9.

10.

11.

12.

TEACHER NOTE: Have students make a figure with a rubber band. Then ask another student to use a rubber band of a different color to make a figure symmetrical to the first.

Congruent figures

Two figures are congruent if they have the same shape and size. They will match when one is placed over the other.

Which figure is congruent to the shaded figure in each row?

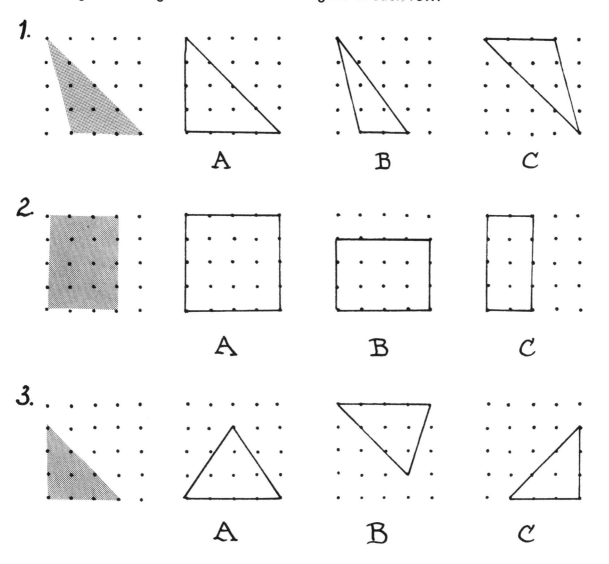

TEACHER NOTE: Have students work together in pairs with geoboard. One student makes a figure with a rubber band. The other student makes a figure congruent to the first figure.

The shapes of geometry can be seen all around us.

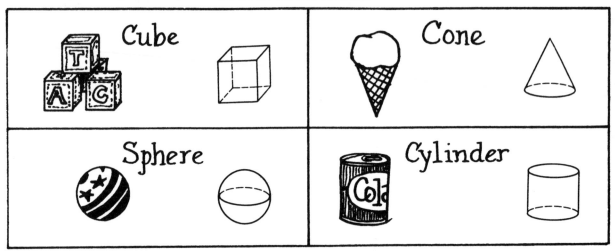

Cube   Cone

Sphere   Cylinder

1. Which of these is a cube?

(A) 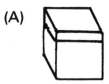   (B)

(C)   (D)

2. Which of these is a cylinder?

(A)    (B)

(C)   (D)

3. Which of these is a cone?

(A)    (B)

(C)    (D)

4. Which of these is a sphere?

(A)   (B)

(C)   (D)

A What shapes make up the sides of a cube? How many of them are there?
  Draw a pattern that would cover all the sides of a cube.

B What shape is the top and bottom of a cylinder? What shape is the side of a cylinder?
  Draw a pattern that would cover the top, side and the bottom of a cylinder.

Name each shape

1.    2.   3.   4.

_____   _____   _____   _____

5.   6.   7.   8.

_____   _____   _____   _____

Which geometric shape most closely resembles

9. a piece of chalk _____   10. the top part of a funnel _____

11. an orange _____   12. a can of soup _____

13. a basketball _____   14. ice from a tray _____

15. Trace the pattern. Cut and fold the pattern on the lines.
What geometric solid is made with this pattern? _____

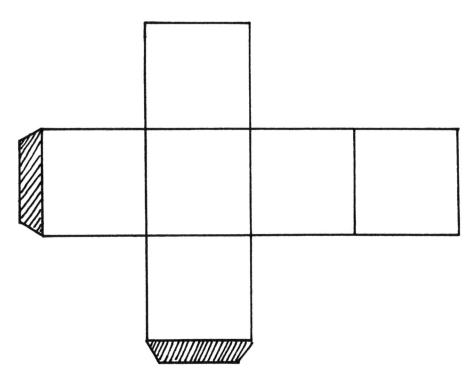

Name _____ Score _____

Geometry Post-Test: Level B

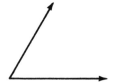

35.

What is the name of this figure?

_____

38. **Which shows a line of symmetry?** ____

(A)   (B)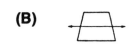

(C)    (D)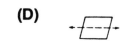

36.   **Line _____ is horizontal.**

(A)   (B)

(C)   (D)

39. **Which figures are congruent?** _____

(A)   (B)

(C)    (D)

37.   **Lines _____ do not intersect.**

(A)    (B)

(C) 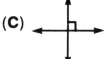   (D)

40.

What is the name of this solid?

_____

STOP

Name _____ Score _____

Measurement Pre-Test: Level B

41. What is the time?

_____

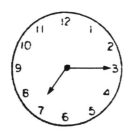

42.

SEPTEMBER

S	M	T	W	T	F	S
					1	2
3	4	5	6	7	8	9
10	11	12	13	14	15	16
17	18	19	20	21	22	23
24	25	26	27	28	29	30

The third Wednesday falls on what date? _____

43.

What is the measure of the line to the nearest ½ inch?

_____

44. 1 yard = _____ feet

1 pound = _____ ounces

1 gallon = _____ quarts

45. 1 meter = _____ centimeters

1 kilogram = _____ grams

1 liter = _____ milliliters

46. What is the perimeter? _____ cm

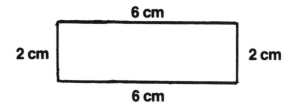

47. You buy a box of mini-doughnuts for $0.39. How much change should you receive from a $1.00 bill?

_____

50.

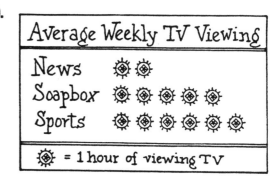

How many more hours are spent per week watching sports than news?

_____

TEACHER NOTE: You may help students read words when requested. Do not explain the meaning of the words.

Telling time to 15 minutes

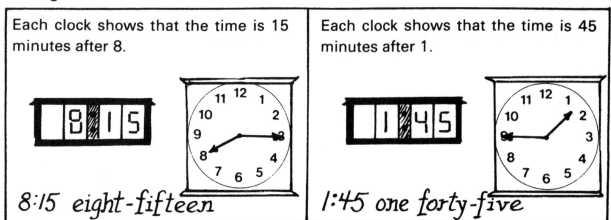

Each clock shows that the time is 15 minutes after 8.

8:15 eight-fifteen

Each clock shows that the time is 45 minutes after 1.

1:45 one forty-five

What time is it?

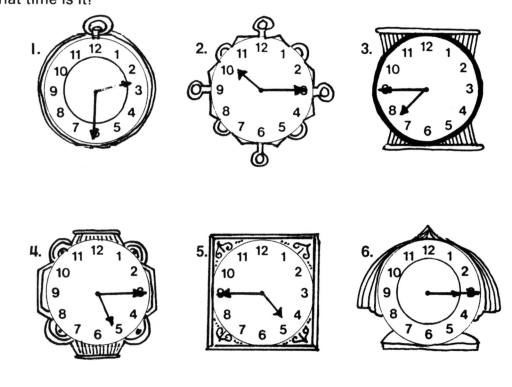

1.

2.

3.

4.

5.

6.

Draw the hands to show the time.

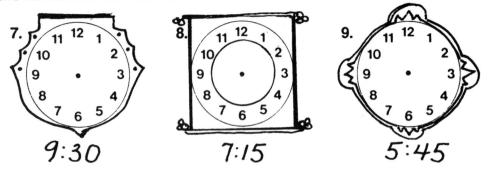

7. 9:30

8. 7:15

9. 5:45

Telling time to 5 minutes

Every space between the numbers around the clock stands for 5 minutes.

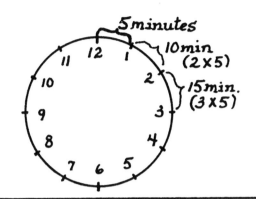

The clock shows that the time is 35 minutes after 2. (The big hand is 7 spaces from the 12, so 7 x 5 = 35).

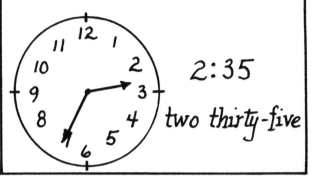

2:35

two thirty-five

Complete the multiplication facts with 5's.

X	1	2	3	4	5	6	7	8	9	10	11	12
5					25					50		

What time is it?

1.

2.

3.

4.

5.

6.

Measuring temperature and weight

We measure temperature with a thermometer marked in degrees (°). The thermometer reads 75 ° F.	We measure weight with a scale. The scale reads 68 pounds.

What is the temperature?

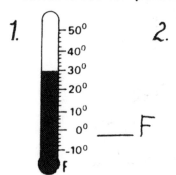

1. ___ F   2. ___ C   3. ___ F   4. ___ C

What is the weight?

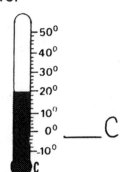 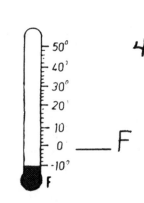 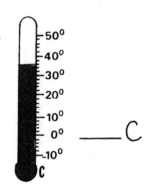

5. _____ pounds   6. _____ kilograms   7. _____ pounds   8. _____ kilograms

(A) Use a Fahrenheit thermometer to find the temperature of:
a) the room
b) a glass of water
c) a cup of hot water

(B) Use a Celsius thermometer to find the temperature of:
a) the room
b) a glass of water
c) a cup of hot water

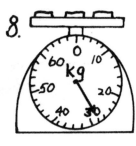

Calendar

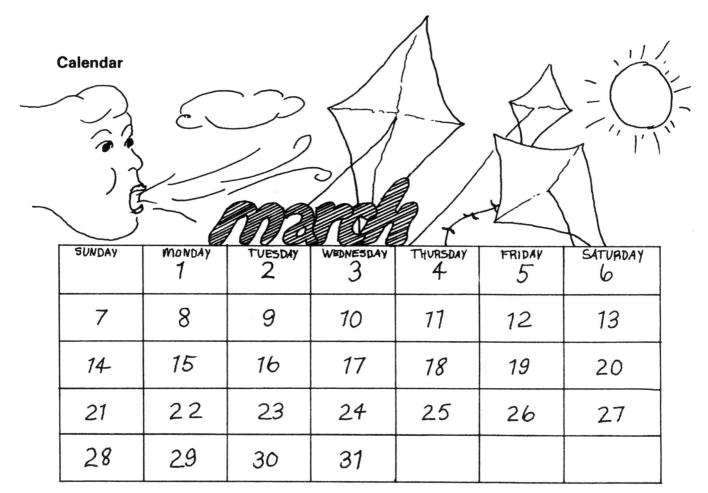

SUNDAY	MONDAY	TUESDAY	WEDNESDAY	THURSDAY	FRIDAY	SATURDAY
	1	2	3	4	5	6
7	8	9	10	11	12	13
14	15	16	17	18	19	20
21	22	23	24	25	26	27
28	29	30	31			

1. How many days are in March? _____

2. On what day of the week does March 1 fall? _____

3. The last of the month is on what day of the week? _____

4. Which dates of the month are on Tuesday? _____

    Which of these Tuesdays comes first? _____

    Which of these Tuesdays comes second? _____

5. On what date does the fourth Thursday fall? _____

6. What is the date a week before March 16? _____

7. What is the date a week after March 23? _____

8. On what day of the week does March 8 fall? _____

9. On what date does the third Sunday fall? _____

10. What is the date two weeks after March 11? _____

Measuring length

The centimeter is a standard unit of measure in the metric system.

The metric system is used in most countries around the world.

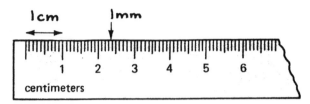

This ruler is marked in units called centimeters (cm) and millimeters (mm).
There are 10 millimeters in 1 centimeter.

Cut out the centimeter ruler on the back inside cover.
Measure each line to the nearest centimeter.

1. _____ _____ cm

2. _____ _____ cm

3. _____ _____ cm

4. _____ _____ cm

5. _____ _____ cm

Measure each line to the nearest half centimeter (.5 cm).

6. ⊢ _____ cm

7. ⊢—⊣ _____ cm

8. ⊢———⊣ _____ cm

9. ⊢————————⊣ _____ cm

10. ⊢———⊣ _____ cm

Draw a line segment with these measures.

11. 5 cm

12. 3 cm

13. 5.5 cm

14. 2.5 cm

TEACHER NOTE: Have students measure ten objects in the room to the nearest centimeter. Put the results in order from least to greatest; then, the objects, to check the order.

The inch is a standard unit of measure in the customary system.

The customary system is often used in the United States and in England.

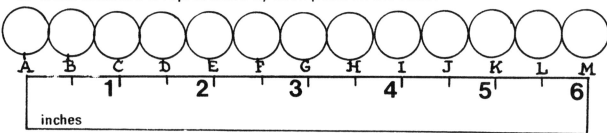

This ruler is marked in units called inches (in.).
The small mark between each number is one-half inch (½ in.).

Cut out the inch ruler on the back inside cover.
Measure each line to the nearest inch.

1. _____ _____ in.

2. _____ _____ in.

3. _____ _____ in.

4. _____ _____ in.

5. What measure is represented by each point on the ruler?

How long is each line to the nearest ½ inch?

6. _____ _____ in.

7. _____ _____ in.

8. ___ _____ in.

9. _____ _____ in.

10. _____ _____ in.

49

Measure each line to the nearest ½ inch.

1. _____ _____in.

2. _____ _____in.

3. _____ _____in.

4. _____ _____in.

5. _____ _____in.

6. _____ _____in.

7. _____ _____in.

8. _____ _____in.

9. _____ _____in.

10. _____ _____in.

11. Skip count by ½'s from 0 to 9½.

  _0_,___,___,___,___,_2½_,___,___,___,___,

  ___,___,___,___,___,___,___,___,___,_9½_.

Draw a line segment with these measures.

12. 2½ inches

13. 5½ inches

14. 4 inches

15. 1½ inches

Customary units of length

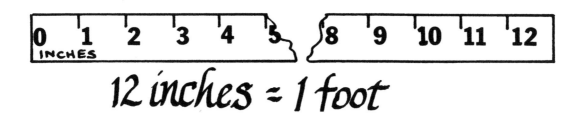

12 inches = 1 foot

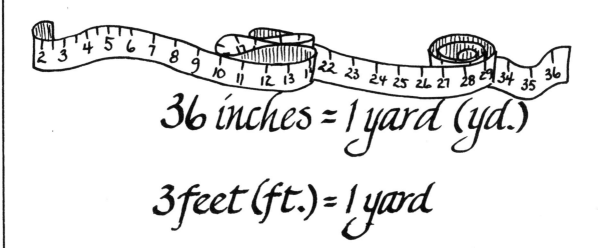

36 inches = 1 yard (yd.)

3 feet (ft.) = 1 yard

(not actual size)

Complete

1. 1 ft. = _____ in.   2. 2 ft. = _____ in.   3. 3 ft. = _____ in.

4. 12 in. = _____ ft.   5. 36 in. = _____ ft.   6. 24 in. = _____ ft.

7. 1 yd. = _____ ft.   8. 2 yd. = _____ ft.   9. 5 yd. = _____ ft.

10. 12 ft. = _____ yd.   11. 9 ft. = _____ yd.   12. 24 ft. = _____ yd.

13. 36 in. = _____ yd.   14. 2 yd. = _____ in.   15. 5 yd. = _____ in.

16. 1 ft. 2 in. = _____ in.   17. 2 ft. 6 in. = _____ in.

18. 15 in. = _____ ft. _____ in.   19. 26 in. = _____ ft. _____ in.

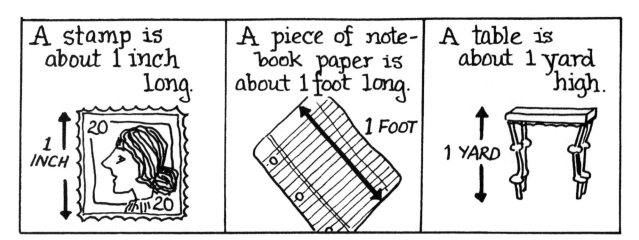

| A stamp is about 1 inch long. | A piece of note-book paper is about 1 foot long. | A table is about 1 yard high. |

Would you measure in inches or feet?

1.  STAPLER

_____

2.

_____

3.

_____

4. length of a room

_____

5. height of a squirrel

_____

6. length of a pencil

_____

Estimate the measure of each object. Then measure each length.

	OBJECT	ESTIMATE	ACTUAL MEASUREMENT
7.	width of your desk	in.	inches
8.	length of your desk	ft.	feet
9.	width of your classroom	ft.	feet
10.	length of your classroom	yd.	yards
11.	height of a windowsill	ft.	feet
12.	height of a door	ft.	feet.
13.	length of your pencil	in.	inches

14. Order the objects (numbered 7-13) from shortest to longest:

<u>13</u> ___ ___ ___ ___ ___ ___
SHORTEST   LONGEST

TEACHER NOTE: Write the actual measurement for each object ordered in #14 above next to the object. Discuss why the measurement numbers do not go in order.

Customary units of capacity

Capacity means how much something holds when filled up.
We often measure capacity in cups (c.), pints (pt.), quarts (qt.) and gallons (gal.).

Would you measure in cups, pints, quarts or gallons?

1. water to bake a cake _____

2. bath tub _____

3. carton of cream _____

4. swimming pool _____

5. a pitcher of lemonade _____

6. a glass of orange juice _____

Complete

7. 1 pt. = _____ c.   8. 1 qt. = _____ pt.   9. 1 gal. = _____ qt.

10. 2 pt. = _____ c.   11. 3 qt. = _____ pt.   12. 4 gal. = _____ qt.

13. 6 c. = _____ pt.   14. 4 pt. = _____ qt.   15. 12 qt. = _____ gal.

16. 1 qt. = _____ c.   17. 2 qt. = _____ c.   18. 16 c. = _____ qt.

Which is greater?

19. 2 pt. or 2 qt.   20. 4 gal. or 4 qt.   21. 2 c. or 2 qt.

22. 6 pt. or 1 qt.   23. 4 c. or 1 pt.   24. 6 pt. or 2 gal.

Customary units of weight

Weight means how heavy something is.
We measure weight in ounces (oz.) and pounds (lb.)

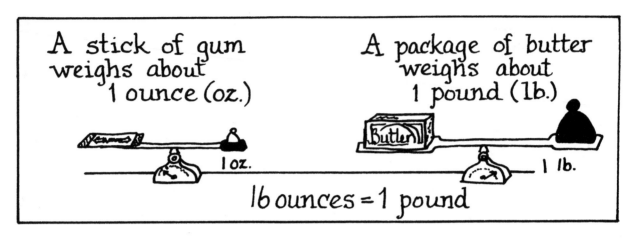

A stick of gum weighs about 1 ounce (oz.)

A package of butter weighs about 1 pound (lb.)

1 oz.

1 lb.

16 ounces = 1 pound

Would you measure the weight in ounces or pounds?

1. an apple _____

2. a dog _____

3. a can of pepper _____

4. a bag of potatoes _____

5. a sack of sugar _____

6. a watch _____

Complete

7.   1 pound = _____ oz.   8.   3 lb. = _____ oz.   9.   4 lb. = _____ oz.

10.   16 oz.   = _____ lb.   11.   48 oz. = _____ lb.   12.   32 oz. = _____ lb.

13. 1 lb. 4 oz. = _____ oz.

14. 2 lb. 6 oz. = _____ oz.

15. 5 lb. 10 oz. = _____ oz.

16. 3 lb. 2 oz. = _____ oz.

17. 20 oz. = _____ lb. _____ oz.

18. 40 oz. = _____ lb. _____ oz.

19. 50 oz. = _____ lb. _____ oz.

20. 160 oz. = _____ lb. _____ oz.

21. 1 foot = _____ inches

22. 1 yard = _____ feet

23. 1 pt. = _____ cups

24. 1 qt. = _____ pt.

25. What units measure length in the customary system? _____, _____, _____.

26. What units measure capacity in the customary system? _____, _____, _____.

_____.

27. What units measure weight in the customary system? _____, _____.

Units of length in the metric system

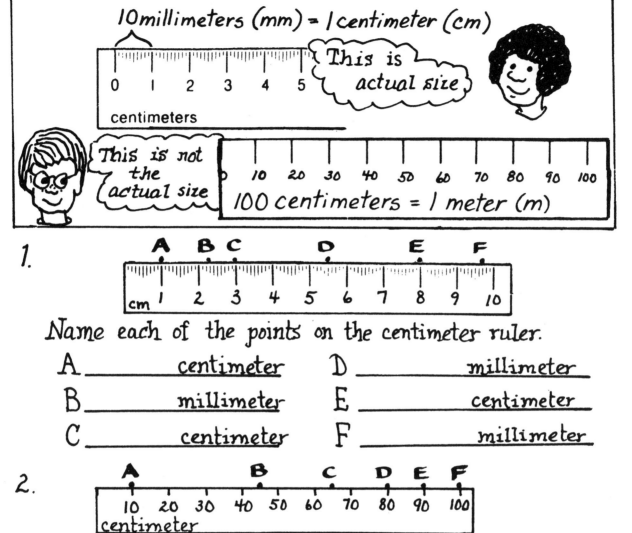

1. Name each of the points on the centimeter ruler.

A _____ centimeter _____   D _____ millimeter _____

B _____ millimeter _____   E _____ centimeter _____

C _____ centimeter _____   F _____ millimeter _____

2. Name each of the points on the meter stick

A _____ centimeters _____   D _____ centimeters _____

B _____ centimeters _____   E _____ centimeters _____

C _____ centimeters _____   F _____ centimeters _____

3. 1m = _____ cm   4. 2m = _____ mm   5. 5m = _____ cm

6. 100cm = _____ m   7. 300cm = _____ m   8. 400cm = _____ m

9. 1cm = _____ mm   10. 2cm = _____ mm   11. 10mm = _____ cm

The thickness of a dime is about 1 millimeter.	The width of your little finger is about 1 centimeter.	The height of your school desk is about 1 meter.
about 1 mm	about 1cm	about 1meter

Would you measure in millimeters, centimeters or meters?

1. _____

2. _____

3. _____

4. a piece of chalk _____

5. thickness of 10 sheets of paper _____

6. width of the hall _____

Estimate the measure of each object. Then measure each with a centimeter ruler or meter stick.

	OBJECT	ESTIMATE	ACTUAL MEASUREMENT
7.	width of your desk	cm	cm
8.	length of chalkboard	m	m
9.	length of a paper clip	mm	mm
10.	height of your desk	m	m
11.	width of your thumb	cm	cm
12.	a new piece of chalk	cm	cm
13.	height of a door	cm	cm

14. Order the objects (numbered 7-13) from shortest to longest:

_____ _____ _____ _____ _____ _____ _____
SHORTEST   LONGEST

Longer lengths in the metric system are measured in
KILOMETERS.

A long pace is about the length of one meter.

1000 paces or 1000 meters = 1 kilometer

The distance between two cities would be measured in kilometers.

BERLIN 38KM

Would you measure in centimeters, meters or kilometers?

1. height of a flag pole _____

2. length of a pen _____

3. car speed per hour _____

4. height of a glass _____

5. length of a marathon run _____

6. length of a swimming pool _____

7. 1 Km = ____ m

8. 2 Km = ____ m

9. 4 Km = ____ m

10. 1000 m = ____ Km

11. 3000 m = ____ Km

12. 6000 m = ____ Km

13. 5500 m = ____ Km ____ m

14. 4300 m = ____ Km ____ m

15. The distance between Springfield and Janesville is 52 Km. How many meters is this? _____

Use a meter stick to find the length of your long pace.

How many paces do you walk in a block?

How many blocks would you walk to have walked a KILOMETER?

Units of capacity in the metric system

We measure capacity in milliliters (mL) and liters (L). Small amounts of liquid are measured in milliliters. Larger amounts are measured in liters.

A cube measuring 1 centimeter on each edge will hold 1 milliliter of water. 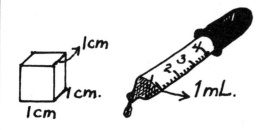	A cube measuring 10 centimeters on each edge will hold 1 liter of water. 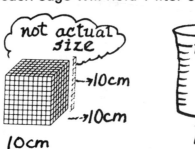

1000 milliliters (mL)= 1 liter (L)

Would you measure how much each container holds in liters or milliliters?

1.  _____

2.  _____

3.  _____

4. The juice of a lemon

5. The gas in a tank

6. Perfume in a bottle

Complete

7. 1 L   = ___ mL   8. 3 L   = ___ mL   9. 2 L   = ___ mL

10. 1000 mL= _____ L   11. 4000 mL= _____ L   12. 5000 mL= _____ L

13. 1 cm   = __ mm   14. 1 m   = ___ cm   15. 200 cm = ___ m

Which measurement is more reasonable?

16. a teaspoon: 4 mL or 4 L

17. a can of paint: 4 mL or 4 L

18. a bath tub: 100 mL or 100L

19. a baby bottle: 150mL or 150L

20. a glass: 300 mL or 300L

21. a metric liter: 1 customary cup or 1 customary quart

Units of weight in the metric system

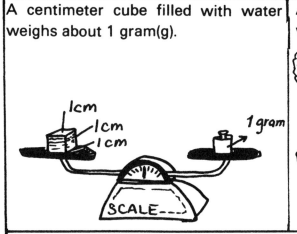

A centimeter cube filled with water weighs about 1 gram(g).

1cm
1cm
1cm

1 gram

SCALE

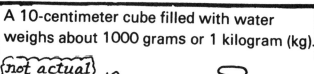

A 10-centimeter cube filled with water weighs about 1000 grams or 1 kilogram (kg).

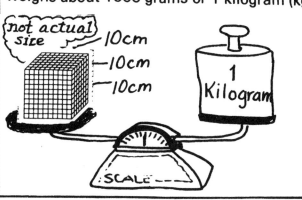

not actual size
10cm
10cm
10cm

1 Kilogram

SCALE

1000 grams (g) = 1 Kilogram (kg)

Would you measure the weight in grams or kilograms?

1.  SOAP

2.

3.

4. A tennis ball

5. Yourself

6. An apple

Complete

7. 1 kg   = _____ g   8. 2 kg   = _____ g   9. 4 kg   = _____ g
10. 1000 g = _____ kg   11. 3000 g = _____ kg   12. 2000 g = _____ kg
13. 1 cm   = __ mm   14. 1 m   = __ cm   15. 1 L   = __ mL

Which measurement is more reasonable?

16. 2 paper clips: 1 gram or 1 kilogram
17. a textbook: 1 gram or 1 kilogram
18. a man: 60 g or 60 kg
19. a quarter: 3 g or kg

20. What units measure length in the metric system? _____ , _____ , _____ , _____ .
21. What units measure capacity in the metric system? _____ , _____ .
22. What units measure weight in the metric system? _____ , _____ .

Find five items in your room and order them from lightest to heaviest. Verify your estimates with a scale.

Problem Solving Steps

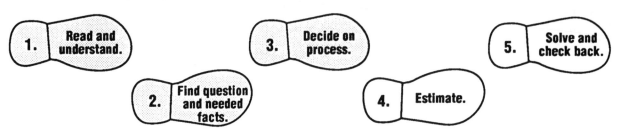

1. Read and understand.

2. Find question and needed facts.

3. Decide on process.

4. Estimate.

5. Solve and check back.

Mr. Romero's class has a guinea pig named Barney for the school year. Answer these story problems about Barney.

1. Barney eats 3 ounces of pellet food each day. How many ounces will he be fed in 174 days of school?	2. Barney's water bottle holds 6 ounces of water. If Barney drank 546 ounces of water, how many bottles of water did he drink?
3. The students will take care of Barney during the summer vacation. There are 27 students, and each student will keep Barney for 3 days. How many days will Barney be visiting students?	4. Barney is 8 inches long and weighs 6 ounces. His cage is 3 times longer than Barney. How long is the cage?
5. When the class bought Barney at the pet store, there were 36 guinea pigs and 27 gerbils. If each guinea pig weighed 9 ounces, how much did they weigh altogether?	6. The children put 6 pounds on one side of the balance scale and Barney on the other side. If Barney weighs 6 ounces, how many <u>more</u> guinea pigs than Barney are needed to balance the scale?

7. Write a story problem about Barney.

_____

_____

Finding perimeter

The distance around the outside of a figure is its perimeter.
One way to find perimeter is to lay units of cubes along the outside edges and count the cubes.

Count and number the cubes
the outside edges.

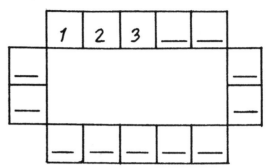

The perimeter of the rectangle
is 14 centimeters.

About how many centimeter blocks fit around the perimeter of each figure?

1.

Perimeter = _____ cm

2.

Perimeter = _____ cm

3.
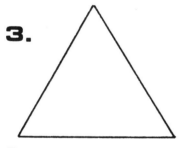
Perimeter = _____ cm

4.
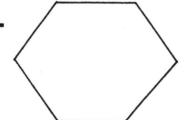
Perimeter = _____ cm

5.

Perimeter = _____ cm

6.
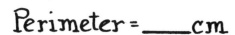
Perimeter = _____ cm

When the sides of a figure are marked off in equal units, we can count the spaces to find the perimeter.

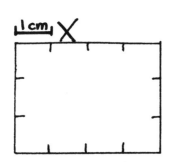

Mark an X on each unit around the figure.
How many X's did you mark?
What is the perimeter of this rectangle?

Find the perimeter of each figure by counting the units on all sides.

1.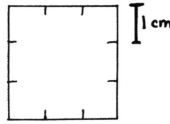

Perimeter = _____ cm

2.

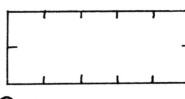

Perimeter = _____ cm

3.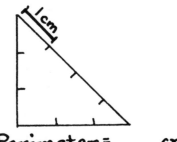

Perimeter = _____ cm

4.

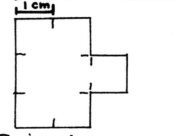

Perimeter = _____ cm

Find the perimeter of each of the ten Cuisenaire rods. Trace the outline of each rod on centimeter graph paper. Count the units around the outside.

Color	Perimeter	Color	Perimeter
white		dk. green	
red		black	
lt. green		brown	
purple		blue	
yellow		orange	

Another way to find perimeter is to measure each side with a ruler. 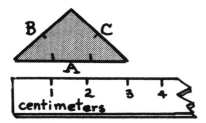	Then... add the measurements to find the perimeter. side A = 3 cm side B = 2 cm side C = 2 cm Perimeter = 3+2+2 = 7cm

Use your centimeter ruler to find the perimeter of each figure.

1.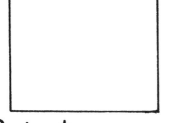

Perimeter = ___ cm

2.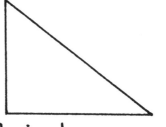

Perimeter = ___ cm

What is the perimeter?

3.

___ cm

4.

___ cm

5.

___ cm

6.

___ cm

7.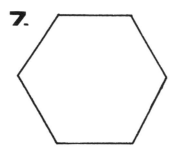

___ cm

8.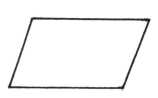

___ cm

Perimeter on dot paper or geoboards

The distance between dots on the paper is 1 centimeter (cm).
Find the perimeter of each figure to the nearest centimeter.

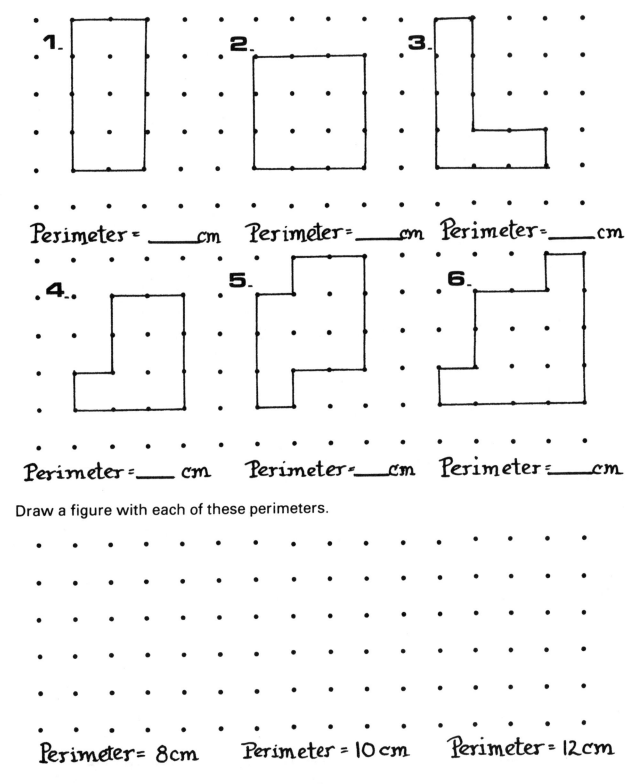

1. Perimeter = _____ cm

2. Perimeter = _____ cm

3. Perimeter = _____ cm

4. Perimeter = _____ cm

5. Perimeter = _____ cm

6. Perimeter = _____ cm

Draw a figure with each of these perimeters.

Perimeter = 8 cm   Perimeter = 10 cm   Perimeter = 12 cm

Area

The number of square units needed to cover the surface of a figure is its area.

Count the number of patches on the quilt.

The area is ____20____ patches

or ____26____ square units.

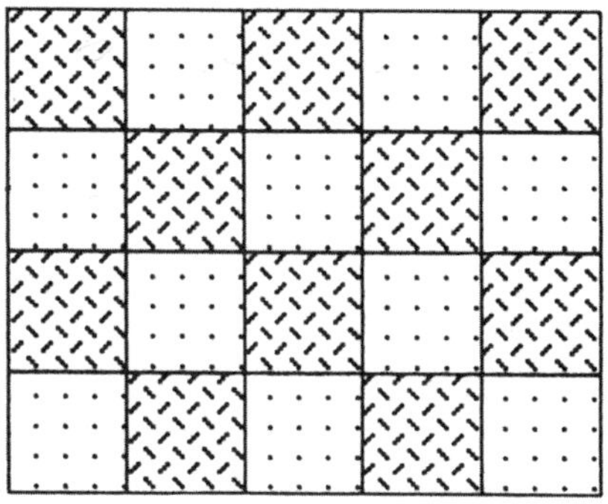

---

1. The diagram shows a square trivet made of tiles. What is the area of the trivet?

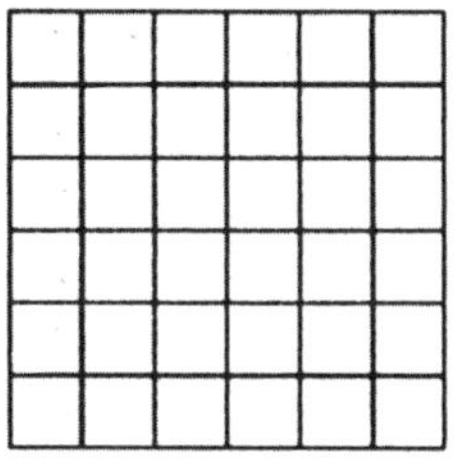

   ____36____ tiles

2. The diagram shows a rectangular trivet made of tiles. What is the area of the trivet?

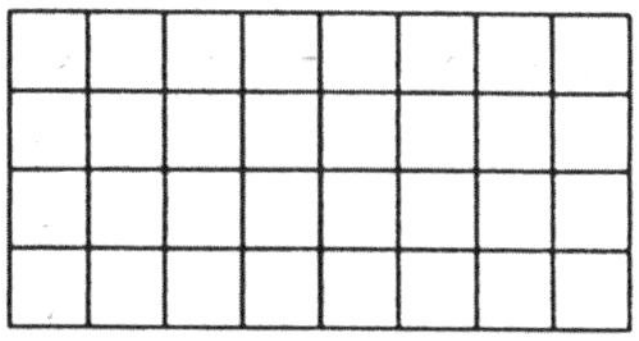

   ____32____ tiles

3. Placemats are made by pasting small squares of different colors on construction paper. Each small square is 1 square inch. What is the area of the place mat?

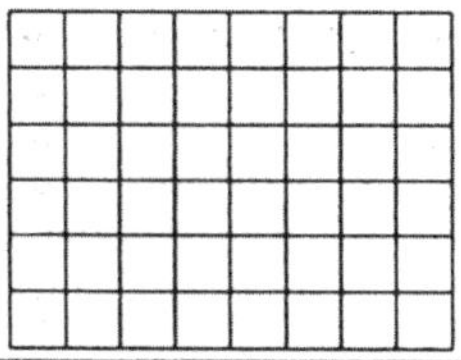

   ____48____ square inches

4. A checker board is made of small square units. What is the area of the checkboard?

   ____64____ square units

5. The diagram shows a kitchen floor. Each tile is 1 square foot. What is the area of the kitchen?

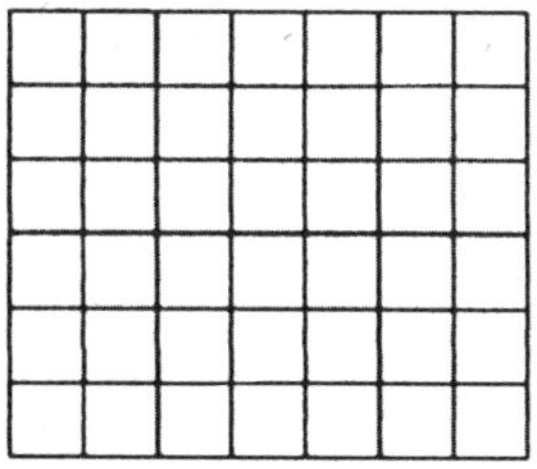

   ____42____ square feet

6. The diagram shows an entry way floor. Each tile is 1 square foot. What is the area of the entry?

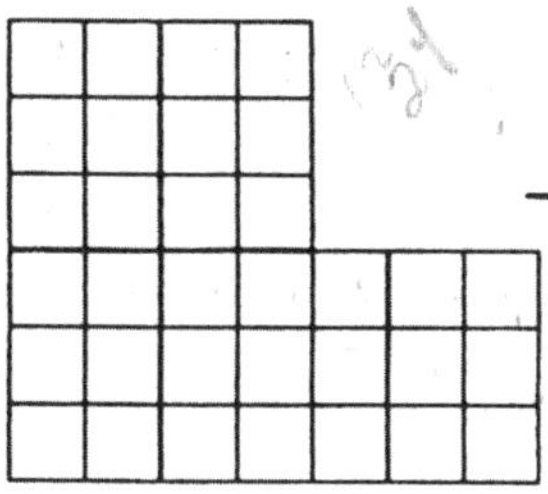

   ____33____ square feet

Estimates of areas

Which is the best estimate of the area of the polygon drawn on the grid?

1. Estimated area = _____ square units.

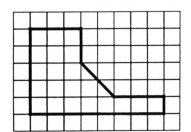

(A) 22 square units
(B) 23 square units
(C) 24 square units
(D) 40 square units

2. Estimated area = _____ square units.

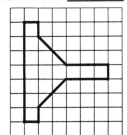

(A) 14 square units
(B) 16 square units
(C) 22 square units
(D) 42 square units

3. Estimated area = _____ square units.

(A) 25 square units
(B) 30 square units
(C) 32 square units
(D) 40 square units

4. Estimated area = _____ square units.

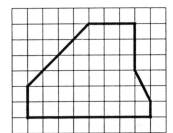

(A) 23 square units
(B) 35 square units
(C) 36 square units
(D) 48 square units

5. Estimated area = _____ square units.

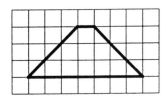

(A) 12 square units
(B) 14 square units
(C) 15 square units
(D) 21 square units

6. Estimated area = _____ square units.

(A) 15 square units
(B) 16 square units
(C) 18 square units
(D) 19 square units

Problem solving: mental math

Fred uses the letters in words to help him remember the meaning of the words. How might Fred remember the meaning of perimeter and area?

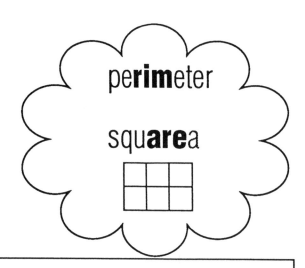

1. How many inches of molding are needed to frame the picture?

   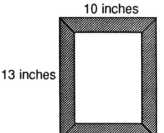

   _____inches

2. A kitchen counter is covered with 1 inch square tiles. What is the area of the counter?

   _____square inches

   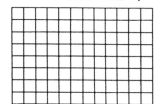

3. Sally is planting a flower bed. The bed is shaped like a triangle. She wants to put a small picket fence around the garden. How many feet of fencing will she need?

   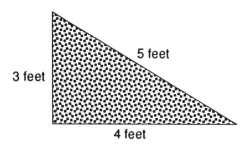

   (A) 7 feet
   (B) 12 feet
   (C) 17 feet
   (D) 20 feet

4. The shaded area is a diagram of the tiles around a wading pool. Which method gives the area of the tiles around the pool?

   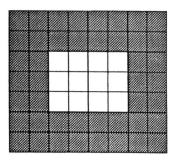

   (A) Count the unshaded squares.
   (B) Count the shaded squares.
   (C) Count all the squares in the diagram.
   (D) Count the units along the outside of the diagram.

Adding and subtracting money

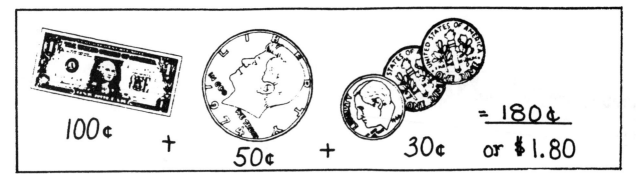

$100¢$ + $50¢$ + $30¢$ = $180¢$ or $\$1.80$

Write the money using the ¢ sign and the $ sign.

1. $1 bill, 1 quarter, 2 dimes = _____ ¢ or $ _____
2. $1 bill, 4 dimes, 1 nickel, 4 pennies = _____ ¢ or $ _____
3. $1 bill, 1 half-dollar, 1 dime, 7 pennies = _____ ¢ or $ _____
4. $5 bill, $1 bill, 3 quarters, 1 dime = _____ ¢ or $ _____

Add the amounts.

5.	$2.45
	  3.10
		6.	$4.75
			  3.25
				7.	$2.65
					  0.86
						8.	$12.45
							  7.25

9.	$0.68
	  0.58
		10.	$1.56
			  0.49
				11.	$0.48
					  0.56
						12.	$1.75
							  2.48

Subtract the amounts.

13.	$3.75
	  1.42
		14.	$4.60
			  2.00
				15.	$6.75
					  0.15
						16.	$3.65
							  1.37

17.	$1.00
	  0.35
		18.	$5.00
			  0.25
				19.	$5.00
					  1.48
						20.	$10.00
							  4.27

What coins are in each bank?

A 26¢ 6 coins   B 77¢ 4 coins   C 48¢ 8 coins   D 42¢ 5 coins

Multiplying with money

MENU

HAMBURGER........80¢
CHEESEBURGER.....90¢
HOT DOG..........40¢
FRENCH FRIES.....30¢
POP..............45¢
MILK.............40¢

Amy bought 6 hamburgers. How much did she spend?

```
  $ 0.80
 x      6
  $ 4.80
```

Use the menu above for problems 1-6.

1. Carrie bought 4 hot dogs. How much for all 4?

2. Song bought 3 french fries. How much for all three?

3. Tad bought 2 hamburgers and 2 milk. What was his total bill?

4. Wendy bought 2 hot dogs, 1 milk, and 1 pop. How much did she spend?

5. Yelena bought 8 pop. How much did she spend?

6. Steven bought 2 hamburgers, 2 french fries and 2 milk. How much did he spend?

7.
```
  $2.15
x    3
```

8.
```
  $0.75
x    5
```

9.
```
  $0.29
x    6
```

10.
```
  $3.72
x    2
```

11.
```
  $2.76
x    4
```

12.
```
  $1.03
x    7
```

13.
```
  $2.99
x    5
```

14.
```
  $4.80
x    8
```

There are 40 possible ways you can make $1.00 with half-dollars, quarters, dimes and nickels. Make a table like this to see if you can find all the ways. Try to use a systematic plan.

	H	Q	D	N
1				
2				
3				
⋮				
40				

Making change

Jacinta bought a cake and a pie. How much change should she get from a $5.00 bill?

COOKIES $.72/bag

PIE $2.35

CAKE $1.90

$1.90 ①
+ 2.35
4.25

$5.00 ②
- 4.25
.75

75¢ change

1. Kelly bought a bag of cookies. How much change did she get from a $1.00 bill?

2. Carlos bought a cake. How much change did he get from a $5.00 bill?

3. Stacy bought a pie. How much change did she get for a $10.00 bill?

4. Paula bought a bag of cookies and a cake. How much change did she get for a $10.00 bill?

5. Seung bought 2 bags of cookies. How much change should he get for a $5.00 bill?

6. Brian bought 2 cakes and a pie. How much change did he get for a $10.00 bill?

7. Carol bought 3 pies and 2 cakes. How much change did she get for a $20.00 bill?

8. Kim bought 2 bags of cookies and a pie. How much change did she get for a $10.00 bill?

9. Afifeh bought a pie and a bag of cookies. How much more did she pay for the pie?

10. Todd bought a bag of cookies and a cake. Noah bought a pie. How much more did Todd spend?

TEACHER NOTE: Have students use play money and practice counting out the change in problems 1-8. In problem 1, say, "72¢, 75¢ (give pennies), 1 dollar (give 1 quarter)."

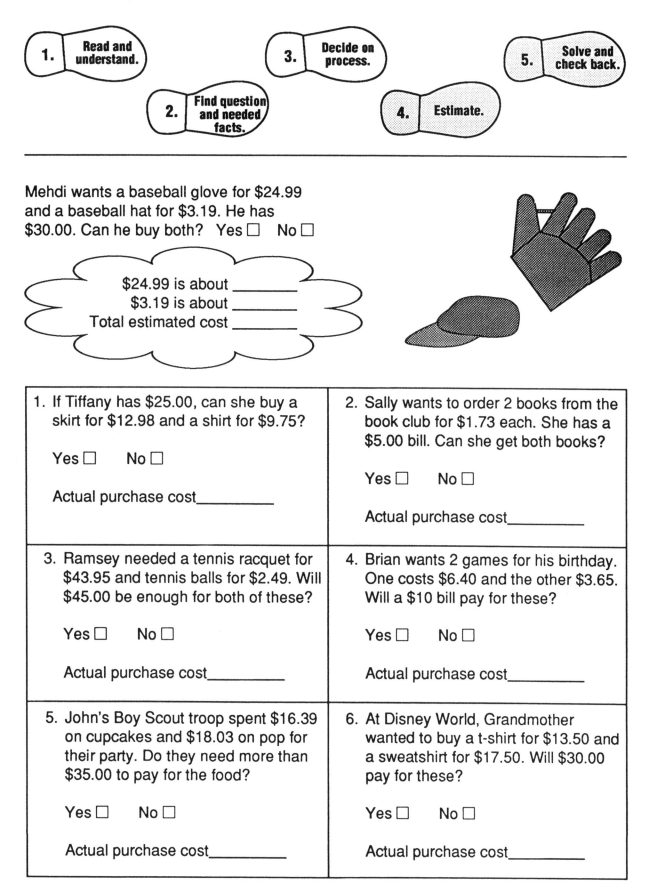

1. Read and understand.

2. Find question and needed facts.

3. Decide on process.

4. Estimate.

5. Solve and check back.

Mehdi wants a baseball glove for $24.99 and a baseball hat for $3.19. He has $30.00. Can he buy both? Yes ☐ No ☐

$24.99 is about _____
$3.19 is about _____
Total estimated cost _____

1. If Tiffany has $25.00, can she buy a skirt for $12.98 and a shirt for $9.75? Yes ☐ No ☐ Actual purchase cost_____	2. Sally wants to order 2 books from the book club for $1.73 each. She has a $5.00 bill. Can she get both books? Yes ☐ No ☐ Actual purchase cost_____
3. Ramsey needed a tennis racquet for $43.95 and tennis balls for $2.49. Will $45.00 be enough for both of these? Yes ☐ No ☐ Actual purchase cost_____	4. Brian wants 2 games for his birthday. One costs $6.40 and the other $3.65. Will a $10 bill pay for these? Yes ☐ No ☐ Actual purchase cost_____
5. John's Boy Scout troop spent $16.39 on cupcakes and $18.03 on pop for their party. Do they need more than $35.00 to pay for the food? Yes ☐ No ☐ Actual purchase cost_____	6. At Disney World, Grandmother wanted to buy a t-shirt for $13.50 and a sweatshirt for $17.50. Will $30.00 pay for these? Yes ☐ No ☐ Actual purchase cost_____

Pictographs

Graphing is one way to record measurements.
A graph with pictures is called a pictograph.

NUMBER OF DIMES EACH STUDENT SAVED

John   ◯ ◯ ◯ ◯ ◯ ◯

Chandra   ◯ ◯ ◯ ◯

Sung   ◯ ◯ ◯

Heather   ◯ ◯ ◯ ◯ ◯

1. How much did John save?

   _____

2. How much did Chandra save?

   _____

3. How much did Sung save?

   _____

4. How much did Heather save?

   _____

5. How much did all four students save?

   _____

6. How much more did John save than Chandra?

   _____

7. Heather wants to save $5.00. How much more must she save?

   _____

8. How much less did Sung save than Heather?

   _____

Study graphs carefully. Read the title and all the words at the bottom and side of the graph.

This graph has a key telling you how much each picture symbol stands for. Find the key and explain it in your own words.

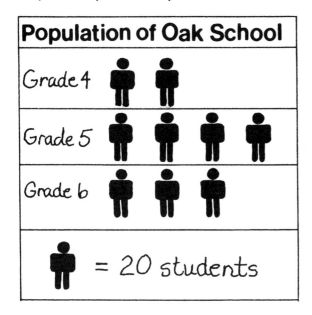

1. How many students in grade 4?

   _____

2. How many students in grade 5?

   _____

3. How many students in grade 4 and grade 5 combined?

   _____

4. How many more students in grade 5 than in grade 4?

   _____

Average Weekly Allowances	
Grade 4	$ $ $
Grade 5	$ $ $ $
Grade 6	$ $ $ $ $ $
$ stands for $1.00	

Number of Bicycles at School	
Fourth	◉ ◉ ◔
Fifth	◉ ◉ ◉ ◉
Sixth	◉ ◉ ◉ ◉ ◉
◉ stands for 10 bicycles	

5. What is the average weekly allowance for fourth graders?

   _____

6. What is the average weekly allowance for a fifth grader?

   _____

7. How much more allowance does the average fifth grader receive than the average fourth grader?

   _____

8. How many bikes are ridden to school by fourth graders?

   _____

9. How many more bikes do sixth graders ride than fifth graders?

   _____

10. How many bikes are ridden by fourth, fifth, and sixth grader altogether?

   _____

Making a bar graph

1. Make a tally showing how many students were born in each month.
    Fill in that number of boxes above each month to make a bar graph.

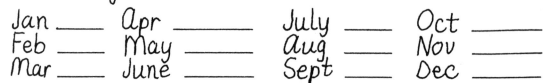

How many were born in each month?

Jan ____   Apr _____   July _____   Oct _____
Feb ____   May _____   Aug _____   Nov _____
Mar ____   June _____   Sept _____   Dec _____

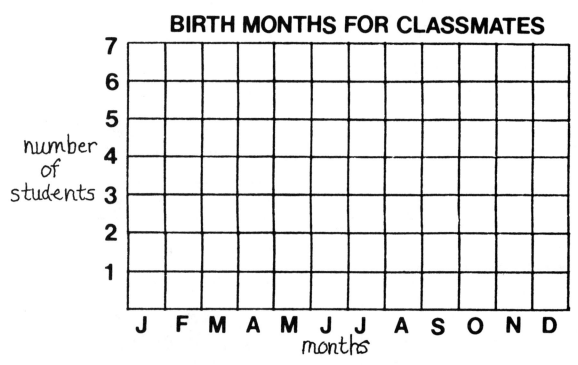

BIRTH MONTHS FOR CLASSMATES

number of students

J F M A M J J A S O N D
months

2. How many students were born in January? _____

3. How many students were born in August? _____

4. In which month were the most students born? _____

5. Is your graph higher in the middle, higher at the ends, or fairly even across the top? Why? _____

TEACHER NOTE: Give each student a unifix cube to help make the tally of birthdays. Write the names of the months across a sheet of paper. Have students, one by one, put the cubes on top of the month in which they were born.

Reading a bar graph

Bar graphs can run in a horizontal or vertical direction.
When the bar doesn't stop right on a line, estimate the number it represents.

1. How many students are 8 years old?

_____

2. About how many are 9 year olds?

_____

3. How many are 10 year olds?

_____

4. How many more 8 year olds are there than 10 year olds?

_____

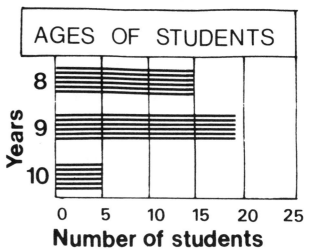

5. What is the favorite subject?

_____

6. What is the least favorite subject?

_____

7. How many chose math as their favorite subject?

_____

8. How many more students prefer math than art?

_____

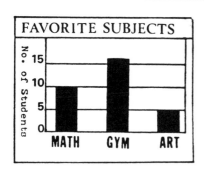

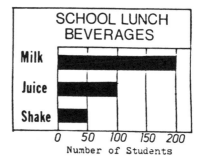

9. What beverage do most students drink at lunch?

_____

10. What is the least favorite beverage?

_____

11. How many more students drink juice than shakes?

_____

12. How many beverages in all were drunk at lunch?

_____

Banana Race

Color Red

Color Yellow

Put two red cubes and one yellow cube in a bag.

Without looking, pick out one cube.

If you pick a red cube, move RED TED one space.

If you pick a yellow cube, move YELLOW FELLOW one space.

Remember to put all the cubes back in your bag before you pick again.

RED TED

YELLOW FELLOW

Which monkey do you think will more likely win the race? _____

Which monkey do you think will less likely win the race? _____

Did you guess right? _____

YELLOW
FELLOW

RED
TED

Color the cubes in the bag red.
Show how many more yellow cubes should be put in each bag.

1. If RED TED and YELLOW FELLOW were equally likely to win.	2. If YELLOW FELLOW was more likely to win than RED TED.
3. If YELLOW FELLOW was less likely to win than RED TED.	4. If YELLOW FELLOW was certain to win every time.
5. If RED TED was certain to win every time.	6. If RED TED and YELLOW FELLOW were equally likely to win.
7. If YELLOW FELLOW was going to win half as often an RED TED.	8. If YELLOW FELLOW was twice as likely to win as RED TED.

Arrangements

Mr. Romero has three reading groups.
He calls them in different orders.

1. Show all the ways he can arrange the order of the three groups.

   $\underline{1}$ $\underline{2}$ $\underline{3}$ ___ ___ ___   ___ ___ ___

   ___ ___ ___   ___ ___ ___

2. If Mr. Romero had only two reading groups, how many ways
   could they be arranged?   _____

3. Sally works in a Submarine sandwich shop. The mini-Sub
   sandwich has lettuce, salami, cheese, and bologne. Show
   the different ways you can arrange the sandwich.
   Use l, s, c and b for lettuce, salami, cheese and bologne.

   $\underline{l}$ $\underline{s}$ $\underline{c}$ $\underline{b}$ ___ ___ ___ ___

   ___ ___ ___ ___   ___ ___ ___ ___

   ___ ___ ___ ___   ___ ___ ___ ___

   ___ ___ ___ ___   ___ ___ ___ ___

   ___ ___ ___ ___   ___ ___ ___ ___

   ___ ___ ___ ___   ___ ___ ___ ___

   ___ ___ ___ ___   ___ ___ ___ ___

   ___ ___ ___ ___   ___ ___ ___ ___

   ___ ___ ___ ___   ___ ___ ___ ___

   ___ ___ ___ ___   ___ ___ ___ ___

   ___ ___ ___ ___   ___ ___ ___ ___

Throw two dice. Shade in a box above each sum. Repeat until a column of boxes is completely shaded in.

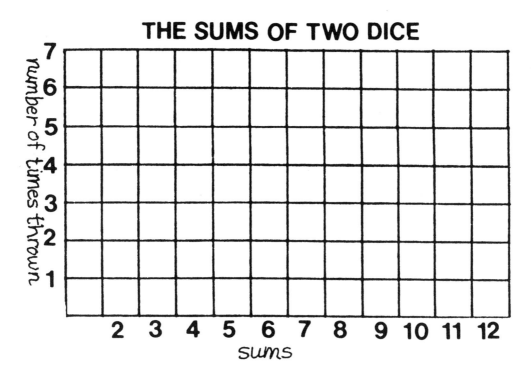

THE SUMS OF TWO DICE

What sum is thrown most often? _____

What sum (or sums) is thrown least often? _____

Is your graph higher in the middle, higher at
the ends or fairly even across the top? _____

Why? _____

 **TEACHER NOTE:** Before doing this exercise, have students make a tally of all the combinations of 1 and 6 which can be thrown to get sums of 2 to 12. "How many ways can you get a sum of 2? (one way: 1,1). "How many ways can you get a sum of 3? (2 ways: 1,2 or 2,1). "How many . . . . .?"

The tally would look like this if a bar graph were made of it.

```
                      (3, 1)
          (2, 1)      (1, 3)
          (1, 2)      (2, 2)
(1,1)
___       ___         ___      ___    ___    ___    ___    ___    ___    ___     ___
 2         3           4        5      6      7      8      9     10     11      12
```

After they have completed the bar graph at the top of the page, compare its shape to the tally of the possible ways. Discuss why the graphs are similar.

Name _____ Score _____

Measurement Post-Test: Level B

41. What is the time?

_____

42. What is the temperature?

_____

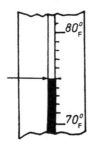

43.

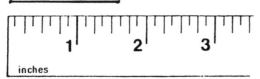

What is the measure of the line to the nearest ½ inch?

_____

44. 1 foot = _____ inches

1 pound = _____ ounces

1 quart = _____ pints

45. 1 kilogram = _____ grams

1 meter = _____ centimeters

1 liter = _____ milliliters

46. What is the perimeter? _____ in.

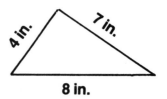

47. You buy a sweatshirt for $7.49. How much change should you receive from a $10.00 bill?

_____

50.

How many more students drink milk at lunch than shakes?

_____

TEACHER NOTE: You may help students read words when requested. Do not explain the meaning of the words.